DARK MATTER
IN THE
UNIVERSE
Second Edition

DARK MATTER
IN THE
UNIVERSE
Second Edition

edited by

John Bahcall
Institute for Advanced Study, Princeton, USA

Tsvi Piran
The Hebrew University, Israel

Steven Weinberg
University of Texas, Austin, USA

 **World Scientific**

NEW JERSEY • LONDON • SINGAPORE • BEIJING • SHANGHAI • HONG KONG • TAIPEI • CHENNAI

Published by

World Scientific Publishing Co. Pte. Ltd.

5 Toh Tuck Link, Singapore 596224

USA office: 27 Warren Street, Suite 401-402, Hackensack, NJ 07601

UK office: 57 Shelton Street, Covent Garden, London WC2H 9HE

British Library Cataloguing-in-Publication Data
A catalogue record for this book is available from the British Library.

DARK MATTER IN THE UNIVERSE (2nd Edition)
4th Jerusalem Winter School for Theoretical Physics Lectures

ISBN-13 978-981-238-840-7
ISBN-10 981-238-840-0

ISBN-13 978-981-238-841-4 (pbk)
ISBN-10 981-238-841-9 (pbk)

Printed in Singapore

FOREWORD TO THE FIRST EDITION

What is the universe made of? We do not know. If standard gravitational theory is correct, then most of the matter in the universe is in an unidentified form that does not emit enough light to have been detected by current instrumentation. Astronomers and physicists are collaborating on analyzing the characteristics of this dark matter and in exploring possible physics or astronomical candidates for the unseen material.

The Fourth Jerusalem Winter School (December 30, 1986 to January 8, 1987) was devoted to a discussion of the so-called "missing-matter" problem. The goal of the School was to make current research work on unseen matter accessible to students or faculty without prior experience in this area. As in previous years, the lectures were informal and the discussions extensive.

The lecturers were J. Bahcall (IAS), R. Blandford (CalTech), M. Milgrom (Weizmann Institute), J. P. Ostriker (Princeton), and S. Tremaine (CITA). Because of the avowedly pedagogical nature of the School and the strong interactions between students and lecturers, the written lectures often contain techniques and explanations that are not available in more formal journal publications. M. Best cheerfully and expertly converted the lectures to their attractive TEX format.

The continued success of the School is made possible by the intelligent and effective leadership of its scientific coordinator, Tsvi Piran, by the strong support of the Israeli Ministry of Science and the Hebrew University, and by Jerusalem's inspiring historical context.

John Bahcall
Co-Director

Steven Weinberg
Permanent Director

CONTENTS

Chapter 7. Gravitational Lenses 103

Roger D. Blandford and Christopher S. Kochanek

Chapter 12.   Dark Matter in Cosmology   **217**

Anthony Aguirre

Chapter 1

INTRODUCTION

John N. Bahcall

School of Natural Sciences,
Institute for Advanced Study,
Princeton, NJ 08540, USA

Every so often in the history of physics a golden opportunity for great progress becomes apparent to contemporary physicists. In the twentieth century, enormous progress was made when, for example, the regularities of atomic spectra became apparent, when the Lamb shift was measured, and when the symmetries of supposedly elementary particles were recognized.

The enigma of dark matter represents a challenge and an opportunity on the same scale as the great physics advances of the twentieth century. Researchers living today are lucky because they can participate in the effort to understand the enigmatic dark matter.

What is the dark matter? It seems very likely that the answer to this question will be of fundamental significance for physics and for astronomy and perhaps for all of science. After all, most of the matter that we know about in the Universe is dark, i.e., we have not been able to detect it yet with our telescopes or other measuring devices. Understanding dark matter will refocus astronomy research and may reveal new types of fundamental particles, e.g., supersymmetric analogues of the known particles.

In 1986, Steven Weinberg and I organized the Fourth Jerusalem Winter School on the subject of dark matter. Tsvi Piran was the very able scientific coordinator. The refereed and edited versions of those pedagogical lectures were published, together with a few relevant reprints, in 1988 by World Scientific Publications as "Dark Matter in the Universe," with J. N. Bahcall, T. Piran, and S. Weinberg as editors. Most of the lectures of that School are reproduced here together with a review article that appears as the last chapter in this book and which summarizes succinctly the state of dark matter research in 2004.

The lectures presented in 1986 represent a solid introduction to the field of dark matter studies (and incidentally to a number of other currently hot topics in astrophysics) and will enable a graduate student or researcher in astronomy or physics to read with understanding the contemporary research papers in the subject.

Most of what we have learned since 1986 is what the dark matter is not; discovering what the dark matter really is remains an exciting challenge.

The lectures and the reprints reproduced in this book have some special advantages for the student and for the active researcher. First of all, the style of the lectures is pedagogical and detailed. This makes it easier to understand the arguments and the assumptions that underlie the conclusions. Second, much of the basic research on dark matter was fresh at the time these lectures were presented and therefore the reader will see the way people participating in these first analyses thought about the puzzles and the challenges. Often, first-look approaches are easier to understand than the more polished presentations that come afterwards.

Of course, much important work has been done in the almost two decades since the Jerusalem Dark Matter Winter School was held. I have resisted the temptation to try to provide a list of references that would cover the more recent literature. It would be a very long list indeed and any given reader would only be interested in a small fraction of the newer literature. Today, there are excellent tools available on the Web so that anyone can find articles that cite earlier articles, like the ones that are cited in the lectures reproduced here. These Web based applications have the advantage of being both convenient and continuously updated.

The most important progress in understanding the astronomical role of dark matter since the 1986 Winter School has been in the context of the formation of large scale structure in the universe and of galaxy formation. The concluding chapter of this book is new; all of the other chapters appeared in the previous edition. The new final chapter, by A. Aguirre, summarizes the current theoretical ideas, the existing observational results, and the future challenges for the very successful cosmological scenario based upon the hypothesis of Cold Dark Matter.

I have chosen to omit from this edition the lectures that I gave at the Winter School on the topic of dark matter in the Disk of our Galaxy. The mathematical techniques described in the omitted lectures have been used by J. Holmberg and C. Flynn (see MNRAS, **313**, 209 (2000) and astro-ph/0405155) to show, using recent Hipparcos observations that are a great improvement over earlier data, that there is no appreciable amount of dark matter in the Galactic Disk. There is a lesson in this development which is useful for all students (and researchers) of astronomy and astrophysics: theoretical inferences are — at best — no better than the observations on which they are based.

Contents of This Book

Chapter 2 of this book is a reprint of a prototypical study of the dark matter in the galaxy NGC 3198, of the kind carried out so fruitfully and systematically by Vera Rubin and her colleagues. The extensive observational data available for this galaxy made possible a detailed analysis that illustrates clearly the astronomical context and the general nature of dark mater in galaxies. This observational paper, by T. S. van Albada and colleagues, symbolizes that the fact that Nature forced us to acknowledge dark matter without the slightest *a priori* theoretical motivation for its existence. Chapter 3, by J. N. Bahcall and S. Casertano, describes one of the most persistent puzzling aspects of dark matter. There is a conspiracy between dark matter and luminous matter to arrange themselves so that the transition in a galaxy from domination by dark matter to domination by luminous matter produces no easily observable features. The reader can see the conspiracy at work in the measurements reported in Chapter 2 for NGC 3198.

Chapter 4, by J. P. Ostriker and C. Thompson, discusses the evolution of globular clusters — systems of typically 10^5 stars. This chapter exposes the readers to some of the classical stellar dynamical questions and techniques that are useful in treating the equilibrium and the evolution of large numbers of stars. The concluding section of this chapter considers the possibility that massive black holes are a significant component of the dark matter in the halos of galaxies like NGC 3198. Chapter 5, also by Ostriker and Thompson, describes in pedagogical detail the effects of positive-energy perturbations in an expanding universe. The authors have in mind perturbations from a galaxy undergoing a burst of star formation, an active quasar, or even a superconducting cosmic string. For the students or researchers interested in astrophysical problems, the most valuable aspect of this chapter is that it describes in an accessible way the "dirty details" of the subject, including hydrodynamics in an expanding universe, shock waves, instabilities, and the influence of dark matter.

Chapter 6, by Scott Tremaine and Hyung Mok Lee, is a concise and accessible introduction to the entire subject of dark matter in galaxies and clusters of galaxies. If you read this pedagogically presented collection of five lectures and verify the equations, then you will be well equipped to do research in the subject. In fact, this chapter is a self-contained summary of the tools needed to address many problems in modern astrophysics. Separate sections are devoted to an overview of the subject of dark matter, the theory of stellar dynamics, the cores of elliptical galaxies and dwarf spheroidal galaxies, the Halo of the Milky Way Galaxy, binary galaxies, and masses of groups and clusters of galaxies. Imagine you were a graduate student in a university where there was no active research in astrophysics. In this case, Chapter 6 would be just what you need. You could easily find a number of interesting and important topics for a graduate thesis by studying this chapter and then applying the techniques described by Tremaine and Lee to new data sets — data sets that are much more extensive than those that were available in 1986. In fact, Chapter 6 is written so clearly and logically that the uninitiated and the expert can both benefit greatly by studying the material systematically.

The ideal way to study dark matter is, in many contexts, by gravitational lensing. Lensing measures the total amount of matter independent of the light that the matter emits. This is just what we want in order to study dark matter. Chapter 7, by R. D. Blandford and C. S. Kochanek, presents an exceptionally clear introduction to the theory and practice of gravitational lensing. The subject has grown enormously since these lectures were given but the principles have not changed. You can find in the lectures everything you need to know to read with understanding the multitude of current papers on gravitational lensing. The topics covered include order of magnitude estimates, the different formalisms for describing gravitational lensing and a comparison of their relative advantages, results for a variety of special cases, generic features of the images, an unusually clear discussion of caustics and catastrophe theory, and compound lenses. Every serious student of dark matter should read this chapter carefully.

Chapters 8–10, by W. H. Press and D. N. Spergel, cover in an introductory but mathematically explicit style three important subjects: inflationary cosmology, cosmic strings, and WIMPS in the Sun and in the laboratory. The lectures are characterized by their directness; the students are given succinct physical arguments followed by the corresponding mathematical equations that in each case summarize the essence of the topic. If you never heard before of inflation or cosmic strings or WIMPS, you could learn what you need to know to be an intelligent consumer of the modern literature on these subjects by reading the introductions in Chapters 8–10. You can't go to a contemporary conference on cosmology without hearing about inflation and you can't go to a contemporary conference on the physics of dark matter without hearing about WIMPS. Although important details have changed since these lectures were written, the basic principles outlined by Press and Spergel are valid today. There is less current interest in cosmic strings today than there was in 1986, largely because they have not been observed, but the subject is still relevant for cosmological investigations.

Nearly all physicists and astronomers assume that dark matter is real. But in the 19th century, nearly all physicists assumed that the aether was real. Consensus does not guarantee correctness. In Chapter 11, Mordehai Milgrom presents a non-relativistic description, usually referred to as MOND, in which the phenomena that are conventionally ascribed to dark matter are instead explained by the failure of Newtonian gravitation at a very low acceleration. This mathematical model makes a number of remarkable predictions. The most remarkable of all the predictions is that every rotation curve of an isolated spiral galaxy can be obtained from the distribution of observed baryonic material using only one parameter (the mass to light ratio). Only a small number of measured rotation curves were available when this prediction was first made in 1983, but today many hundreds of rotation curves are known with good accuracy. If there are any exceptions to the prediction, for the rotation curves of isolated galaxies, of the modified Newtonian dynamics advocated by Milgrom, then they are rare. This is an extraordinary situation. If the usual dark matter picture is correct, then there is no reason why rotation curves measured in different galaxies should not have large random differences. A conspiracy of the kind described in Chapter 3 is required to suppress variety and to hide the signature of dark matter in galactic rotation curves. Something deep is right about MOND, if only to describe in a succinct way a number of *a priori* surprising regularities in the data for galactic systems. Everyone interested in dark matter should read Chapter 11 and think about its implications.

The concluding Chapter 12 by Anthony Aguirre presents the theoretical basis for understanding the role of Cold Dark Matter in determining the observed anisotropies of the Cosmic Microwave Background, the power spectra of the Ly-α forest, and the distribution of galaxies. Aguirre also clearly and succinctly summarizes the theoretical ideas and the observational data related to the formation of galaxies and their halos. In the concluding section of this chapter, Aguirre steps back and outlines objectively and insightfully the current status of galaxy formation theory, as well as the outstanding challenges to the hypothesis of Cold Dark Matter. Everyone interested in modern cosmology can benefit from reading this chapter.

May 2004

Chapter 2

DISTRIBUTION OF DARK MATTER IN THE SPIRAL GALAXY NGC 3198[*]

T. S. van Albada, K. Begeman and R. Sanscisi

Kapteyn Astronomical Institute,
Postbus 800, 9700 AV Groningen, The Netherlands

J. N. Bahcall

Institute for Advanced Study,
Princeton, NJ 08540, USA

Two-component mass models, consisting of an exponential disk and a spherical halo, are constructed to fit a newly determined rotation curve of NGC 3198 that extends to 11 disk scale lengths. The amount of dark matter inside the last point of the rotation curve, at 30 kpc, is at least 4 times larger than the amount of visible matter, with $(M/L_B)_{tot} = 18 M_\odot/L_{B\odot}$. The maximum mass-to-light ratio for the disk is $M/L_B = 3.6$. The available data cannot discriminate between disk models with low M/L and high M/L, but we present arguments which suggest that the true mass-to-light ratio of the disk is close to the maximum computed value. The core radius of the distribution of dark matter is found to satisfy $1.7 < R_{core} < 12.5$ kpc.

1. Introduction

The problem of dark matter surrounding spiral galaxies, made evident by the flatness of rotation curves, is one of the most enigmatic questions in present-day astrophysics. A number of years of intensive research have brought little or no clarification, and suggestions offered in the explanation of flat rotation curves include extremes like the possibility that Newtonian dynamics is at fault (Milgrom 1983; Sanders 1984) and Kalnajs's (1983) conclusion that — for some data sets — there is no problem at all.

In order to set the stage for the discussion in this paper, and to elucidate the confusing situation noted above, it is useful to review what one expects for the shape of the rotation curve. Let us assume that a typical spiral galaxy consists of two distinct distributions of matter: an exponential disk and a de Vaucouleurs spheroid, each with constant M/L. Surface brightness distributions of about half of the galaxies surveyed so far can indeed be explained by the sum of two such

---

[*]Published in *Astrophy. J.* **295**, pp. 305–313, 1985.

components (Boroson 1981; Wevers 1984). Following Boroson, we further assume
that deviations from this simple picture shown by other galaxies can be attributed
to irregularities in the distribution of bright young stars contributing little to the
total mass. The maxima of the rotation curves for spheroid and disk lie at 0.3
effective radii and at 2.2 disk scale lengths, respectively. In the interval between
these two points the shape of the rotation curve depends on the ratio of spheroid to
disk mass, but it can be quite flat. Beyond about 3 disk scale lengths the rotation
curve will show an approximately Keplerian decline.

   In comparing this model with observations we first consider optical data. Rubin
et al.'s rotation curves typically extend to 0.8 times the radius, R_{25}, of the 25th
mag arcsec^{-2} isophote (Rubin 1983). Since the central surface brightnesses of disks
of spiral galaxies lie in the range $20.5 < B_c(0) < 23.0$ (Boroson 1981; Wevers 1984),
optical velocity information stops at 1.5–3.5 disk scale lengths. At 3.5 scale lengths
the rotation curve of an exponential disk has decreased only 8% cent relative to
the maximum value (and even less for a truncated disk; Casertano 1983). With
optical data alone, it is not easy to see the Keplerian decline of the rotation curve.
In many cases, a combined model with a bulge and a disk can produce a circular
velocity that is nearly independent of radius, without the need for nonluminous
matter.

   Another important aspect of this problem is that Rubin et al. (1982; see also
Rubin 1983) find a well-established change in the shape of the rotation curve with
luminosity for Sb galaxies. High-luminosity Sb galaxies show a rapid rise of rotation velocity, V_{cir}, with fractional radius and then reach a more or less constant
level, while rotation curves of low-luminosity Sb galaxies show a more gradual
rise. [Note that the scaling of radius in terms of R_{25} corresponds to expressing
the radius in the number of disk scale lengths if there is no scatter in $B_c(0)$.]
On the bulge disk picture, this progression in shape with luminosity could only
be explained if there exists a strong correlation of bulge-to-disk ratio with luminosity *within* a given Hubble type. The sense required is that high-luminosity
galaxies have a prominent bulge — in terms of mass — while low-luminosity
galaxies have no bulge at all. We show for illustrative purposes in Fig. 1 the
rotation curve for a high-luminosity galaxy, with spheroid and disk parameters
adjusted in such a way that the rotation curve is flat for a large range in radii.
Rotation curves *and* surface photometry are necessary to test whether bulge and
disk properties deduced from the photometry are consistent with the assumption that such a bulge and disk, with constant M/L, produce the rotation curve
observed.

   We conclude from the above discussion that one can explain the shape of rotation
curves inside 3 disk scale lengths, at least in principle, by a combination of matter
distributions with constant M/L, without an additional component of dark matter.
In this respect our conclusions agree with those of Kalnajs (1983). However, rotation
curves obtained from the 21 cm line of neutral hydrogen gas in the outer region of

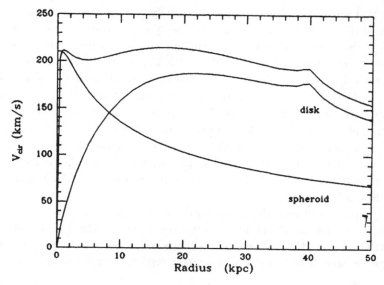

Fig. 1.　Rotation curve due to the sum of a de Vaucouleurs spheroid and a truncated exponential disk. The parameters have been chosen such that the rotation curve is flat over a large range in radii. The peak of the rotation curve of the spheroid lies at 0.8 kpc. Rotation curves such as these can probably explain observed flat rotation curves inside the luminous body of a galaxy, without the need for an additional component of dark matter. R_{eff}(spheroid) = 2.7 kpc; scale length disk, $h = 10$ kpc; truncation radius disk, $R_{\mathrm{trunc}} = 4h$; mass ratio $M_{\mathrm{sph}}/M_{\mathrm{disk}} = 0.25$.

spiral galaxies leave no doubt that "dark halos" do exist[a] (Faber and Gallagher 1979; Bosma 1981). Thus one must have both optical (for the inner regions) and H I data (for the outer regions) to obtain a complete picture of the distribution of matter.

Unfortunately, the distribution of neutral hydrogen in the outer regions of galaxies is sometimes irregular, and, in galaxies seen edge-on, it is not always clear whether H I gas is present at the line of nodes. In such cases, a precise measurement of the circular velocity is not possible. But there are also several galaxies with warped H I disks which can be represented successfully by a tilted ring model. For moderate warps such a tilted ring model should be good enough to allow a reliable determination of the circular velocity.

These considerations show that galaxies with large, relatively unperturbed hydrogen disks, seen at inclinations of say 50° to 80°, are required for studies of the distribution of dark matter. A good example meeting these criteria is NGC 3198. This Sc galaxy, which has no nearby bright companions, has been observed by

---

[a] As pointed out by Freeman (1970) in an epochal paper, rotational velocities of H I (for NGC 300) do not show the expected decline at large radii (that is, beyond the turnover point of the disk). Freeman concluded: "If the H I rotation curve is correct, then there must be undetected matter beyond the optical extent of NGC 300; its mass must be at least of the same order as the mass of the detected galaxy."

Bosma (1981) with the WSRT. Its velocity field is regular and agrees with that for a disk in differential rotation. In addition, the H I distribution extends to at least $2.3R_{25}$, even though Bosma's observations were not particularly sensitive. New 21 cm line observations with the WSRT, with improved sensitivity, have recently been obtained by one of us (Begeman, in preparation). The rms noise in these new H I maps is a factor of 4 smaller than in Bosma's maps, and the velocity field of H I can now be determined out to $2.7R_{25}$ (1.9 Holmberg radii), which corresponds to 11 scale lengths of the disk (see below). Even at these large distances from the center, deviations from axial symmetry in the velocity field are small. To our knowledge, this is the largest number of disk scale lengths over which a galactic rotation curve has been measured. A map of H I contours superposed on a IIIa-J photograph is shown in Fig. 2 (Plate 13).

In this paper we discuss the implications of these observations, in combination with photometry, for the distribution of matter in NGC 3198. Surface photometry of NGC 3198 (Wevers 1984) shows that a single component, i.e. an exponential disk with a scale length of 60″, gives an adequate representation of the distribution of

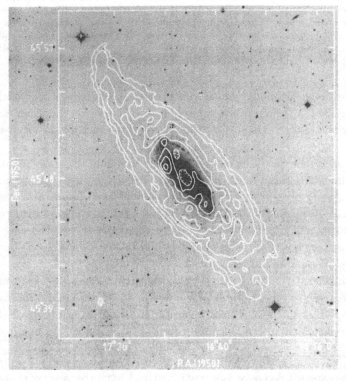

Fig. 2. Full resolution ($25'' \times 35''$) map of the H I column density distribution of NGC 3198 superposed on an optical image of the galaxy (IIIa-J) (with Wr 2c), kindly made available by J. W. Sulentic. Contour levels are 1, 4, 8, 12, ..., 28 times 10^{20} atoms cm^{-2}.

light. Since the circular velocity is essentially constant beyond 2.5 disk scale lengths, this implies that a mass distribution with constant M/L is ruled out.

We find that the amount of dark matter associated with NGC 3198 inside 30 kpc, that is, the outermost point of the rotation curve, is at least 4 times larger than the visible mass. Inside one Holmberg radius (15.9 kpc) the ratio of halo mass to maximum disk mass is 1.5.

The organization of the paper is as follows. In Secs. 2 and 3 we describe the distribution of light and the rotation curve. Two-component mass models fitted to the observed rotation curve are presented in Sec. 4; the results are discussed in Sec. 5.

2. Distribution of Light

Surface photometry of NGC 3198 in three colors (U', $\lambda \approx 3760$ Å; J, $\lambda \approx 4700$ Å; F, $\lambda \approx 6400$ Å) has been published by Wevers (1984). An earlier study of van der Kruit (1979) gives photometry in the J band. Although the photograph of NGC 3198 in Bosma (1981) shows a distinct nucleus, Wevers's radial luminosity profiles can in first approximation be fitted with an exponential disk. Fitting straight lines to these profiles by eye, we find the following scale lengths: U', $63''$; J, $58''$; F, $54''$, with uncertainties of 5%.

Comparison of the U' and F profiles shows that, near the center, there is a distinct color gradient, with the central region being redder. This may be related to the presence of a bulge or a depletion of young stars. For the purposes of this paper, an exponential law is a satisfactory representation of the light distribution; we adopt a scale length of $60''$, corresponding to 2.68 kpc for $H_0 = 75$ km s^{-1} Mpc^{-1}. (From the rotation curve we also find that the bulge must be small; see Sec. 4). A summary of properties of NGC 3198 relevant for this paper is given in Table 1.

Table 1. Properties of NGC 3198

Parameter	Value	Notes
Type	Sc(rs)I–II	1
$V_{\rm hel}$ (km s^{-1})	660 ± 1	2
Distance (Mpc)	9.2	3
B_T^0	10.45	4
$M_{B_T}^0$	-19.36	
$(B-V)_T^0$	0.42	4
$L_B/L_{B_\odot}$	$(8.6 \pm 0.9) \times 10^9$	
$L_V/L_{V_\odot}$	$(7.0 \pm 0.7) \times 10^9$	
Scalelength disk	$60''(= 2.68\,{\rm kpc})$	
R_{25}	$4'.2$	4
Holmberg diameter	$11'.9 \times 4'.9$	5

Notes. — (1) Sandage and Tammann (1981). (2) Begeman (1985). (3) $H_0 = 75$ km s^{-1}Mpc^{-1}. (4) Radius 25th mag arcsec^{-2} isophote; de Vaucouleurs, de Vaucouleurs, and Corwin (1976). (5) Holmberg (1958).

3. Rotation Curve

A rotation curve was derived from the new WSRT observations (FWHM beam 30″)
as follows. (Full details are given elsewhere; Begeman, in preparation). We represent
the hydrogen disk by a number of circular rings, each ring being characterized by an
inclination i, a position angle φ, and a circular velocity V_{cir}. The width of the rings
is 30″ on the major axis. Excluding a sector with opening angle 90° (in the plane
of the galaxy) about the minor axis, we use the grid points at which the velocity
field is recorded inside each ring to obtain a least-squares solution for i, φ, and V_{cir}
as a function of radius. An advantage of this method is that ring-to-ring variations
in i and φ, in combination with the formal errors from the least-squares solution,
indicate to what extent the hydrogen disk is warped. As expected, the position
angle is well determined, while V_{cir} and i are not completely independent. Yet,
taking the dependence into account, the formal errors in V_{cir} and i are extremely
small: $1 \,\mathrm{km\ s^{-1}}$ and 1°, respectively. We find small variations of inclination with
radius: from 72° inside 2′, through a minimum of 70° at 6′, to 76° at 10′ from
the center. There is also a systematic variation of a few degrees in position angle
with radius. Thus, there appears to be a small warp of the disk; it corresponds to
a vertical displacement of 2.3 kpc at the "edge" (29 kpc from the center).

As a further check of these findings, the same procedure was applied to the
northern and southern halves of the galaxy separately. The results clearly show the
symmetric large-scale structure of NGC 3198: both halves show the same dependence of position angle and inclination on radius, and in the region beyond 6′ from
the center, i.e. the interval that is most critical for the subject of this paper, the
inclinations derived separately for the two halves agree to within 1° (see Fig. 3). On
the other hand, the rotation curves for the two halves are slightly different: in the
southern half there is almost no change in V_{cir} with radius beyond the maximum at
3′, but in the northern half V_{cir} decreases slowly between 3′ and 8′ by a few $\mathrm{km\ s^{-1}}$
and then rises between 8′ and 11′. The maximum difference between the two halves
is $6 \,\mathrm{km\ s^{-1}}$.

In the inner region, i.e. inside $\sim$3′, this method of deriving the circular velocity
does not work since there are only a few grid points per ring. Moreover, the gradient
in the velocity field across a ring, and across the beam, must be taken into account.
As described by Begeman, for small radii the circular velocity can be derived from
an l–v diagram along an adopted direction for the major axis and with an adopted
value for the inclination of the disk (l is the position along the major axis). Using
the information in the l–v diagram a first estimate of the rotation curve is obtained
by plotting the velocities corresponding to peak intensity, corrected for inclination,
against l. A correction for beam smearing is then calculated by taking a model velocity field and convolving it with the WSRT beam. (A correction for beam smearing
was calculated and applied for all radii, but beyond 3′ this correction becomes negligibly small.) The rotation curve determined by Cheriguene (1975) from the motion
of H II regions 2′ from the center agrees well with the l–v diagram and H I at the

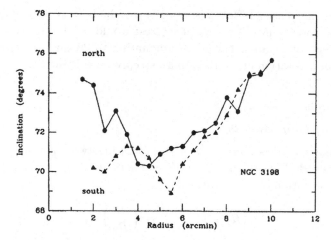

Fig. 3.   Variation of inclination angle with radius in NGC 3198, according to a tilted ring fitted to the velocity field, separately for the northern and southern half of the galaxy. 1′ corresponds to 2.68 kpc.

Table 2.   H I Rotation Curve of NGC 3198.

Distance from Center	V_{hel} (km s^{-1})	Distance from Center	V_{hel} (km s^{-1})
0.25	55 ± 8	4′.5	153 ± 2
0.50	92 ± 8	5.0	154 ± 2
0.75	110 ± 6	5.5	153 ± 2
1.00	123 ± 5	6.0	150 ± 2
1.25	134 ± 4	6.5	149 ± 2
1.50	142 ± 4	7.0	148 ± 2
1.75	145 ± 3	7.5	146 ± 2
2.00	147 ± 3	8.0	147 ± 2
2.25	148 ± 3	8.5	148 ± 2
2.50	152 ± 2	9.0	148 ± 2
2.75	155 ± 2	9.5	149 ± 2
3.00	156 ± 2	10.0	150 ± 2
3.50	157 ± 2	10.5	150 ± 3
4.00	153 ± 2	11.0	149 ± 3

position angle of her data for $r \geq 60''$. At $30''$ from the center, the beam-corrected H I rotation velocities are ~ 20 km s^{-1} higher than those of Cheriquene. H I emission on the scale of the beam ($30''$) has been detected everywhere in the central region.

The final rotation curve is given in Table 2. Inside 2′.5 they correspond to 0.5 times the mean difference between the rotation curves derived separately for the northern and southern halves (~ 2 km s^{-1}). The last two points of the rotation curve are based on adopted values for position angle and inclination.

Between 5′ and 9′, Bosma's (1981) rotation velocities are systematically lower than the new ones by ~ 7 km s^{-1}. This is partly due to the use of a fixed inclination as a function of radius by Bosma. The cause of the remaining difference is not entirely clear. The sense of Bosma's velocity residuals is such that a somewhat larger rotation velocity in the region of interest would have improved the fit. Due to the correction for beam-smearing — amounting to $23\,\mathrm{km\ s^{-1}}$ at $30''$ from the center — the new rotation curve rises more steeply than Bosma's.

4. Mass Models

4.1. *Choice of Components*

We consider mass models consisting of the following two components: (i) a thin exponential disk (de Vaucouleurs 1959; Freeman 1970); (ii) a spherical halo, representing the distribution of dark matter:

$$\rho_{\mathrm{halo}}(R) \propto \left[\left(\frac{a}{R_0} \right)^\gamma + \left(\frac{R}{R_0} \right)^\gamma \right]^{-1}, \tag{1}$$

where R_0 is a fiducial radius (Bahcall, Schmidt, and Soneira 1982; hereafter BSS). The parameter a, which is linked to the core radius, and the exponent γ can be varied freely. We choose R_0 equal to $8\,\mathrm{kpc}$. This allows a comparison of $\rho_{\mathrm{halo}}(R_0)$ for the models with the halo mass density in the solar neighborhood ($\sim 0.01 - 0.12\,M_\odot\,\mathrm{pc}^{-3}$; Bahcall and Soneira 1980). Equation (1) is equivalent to

$$\rho_{\mathrm{halo}}(R) = \rho_{\mathrm{halo}}(0) \left[1 + \left(\frac{R}{a} \right)^\gamma \right]^{-1}.$$

In addition, we estimate the maximum mass of a bulge component, represented by a de Vaucouleurs spheroid, from the shape of the rotation curve in the inner region. To model the spheroid we use the approximations given by Young (1976; see BSS, Table 1).

4.2. *Fits with Exponential Disk and Halo*

To limit the number of free parameters of the models we shall only consider exponential disks with scale lengths equal to that of the light distribution.[b] This is equivalent to the use of exponential disks with M/L independent of radius. As described in Sec. 2, there is no clear evidence for a bulge in NGC 3198. Therefore we will first restrict ourselves to the combination of a disk and halo. The distribution of light also

---

[b]Following a suggestion by J. P. Ostriker and the referee, we have checked the exponential disk approximation of the light distribution by calculating rotation curves directly from the light profiles (assuming an infinitely thin disk). The rate of decline beyond the maximum agrees well with that for an exponential disk with scale length $60''$ for all three curves (U', J, and F). In the inner region there is also good agreement between the exponential disk rotation curve and the rotation curve calculated from the F profile, which presumably gives the better indication of the underlying mass distribution. The U' rotation curve rises less steeply, indicating that compared to F, the dominant population contributing to U' has a central depression in its radial distribution. No J data are available in the inner region. These results confirm our choice of $60''$ for the scale length of the disk. We thank S. Casertano for lending us his programs to perform these calculations.

shows that any truncation of the disk must occur outside 5.5 scale lengths. For such a large truncation radius the difference between a truncated and an untruncated disk is very small as far as the rotation curve is concerned (Casertano 1983; Bahcall 1983). It will therefore be sufficient to consider only the simple case of infinite disks. Another reason for considering infinite disks only is that in the outer regions the contribution of H I to the total mass density cannot be neglected. Between $5'.5$ and $11'$ the surface density of H I is well represented by an exponential with scale length $3'.3$. For the maximum disk case to be discussed below the surface densities of stars and H I are about equal at $R = 5'.5$. The total amount of H I present in NGC 3198 is $4.8 \times 10^9 \, M_\odot$; its contribution to the maximum disk mass is 15%.

Our first model consists of a disk with the largest possible mass and a halo. A strict upper limit for $V_{\max}$ of this disk is 150 km s^{-1}. This choice would require a halo with a hollow core, however, which is implausible. Thus $V_{\max}$ must be somewhat smaller. We find that a reduction of $V_{\max}$(disk) to 140 km s^{-1} is sufficient to allow a halo with a density that decreases monotonically with galactocentric distance. The fit of this disk (total mass $3.1 \times 10^{10} \, M_\odot$) and halo to the observed rotation curve is shown in Fig. 4. The parameters deduced for the halo are not unique; curves with two free parameters are generally sufficient for mass modeling purposes (Kormendy 1982). In this case the halo exponent γ and the scale length a can be varied in a correlated fashion ($1.9 < \gamma < 2.9$, $7 < a < 12$), while $\rho(R_0)$ is fixed to within a narrow range. This freedom in the choice of halo parameters need not concern us: our main interest is the mass distribution in the dark halo, which follows directly from

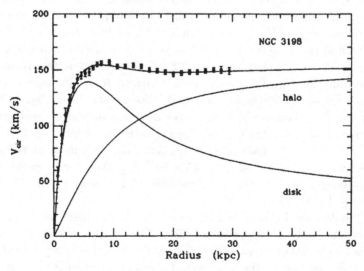

Fig. 4. Fit of exponential disk with maximum mass and halo to observed rotation curve (*dots with error bars*). The scale length of the disk has been taken equal to that of the light distribution ($60''$, corresponding to 2.68 kpc). The halo curve is based on Eq. (1), $a = 8.5$ kpc, $\gamma = 2.1$, $\rho(R_0) = 0.0040 \, M_\odot$ pc^{-3}.

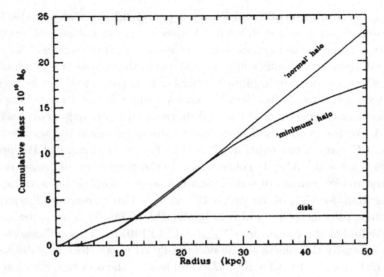

Fig. 5.   Cumulative distribution of mass with radius for disk and halo for the maximum disk mass
case. Two halo fits are shown. The curve labeled "normal" halo is based on Eq. (1); the parameters
of the fit are the same as those in Fig. 4. The curve labeled "minimum" halo is based on Eq. (2);
it corresponds to a density distribution whose slope changes from -2 in the inner regions to -3.5
in the outer regions. This curve represents an estimate of the minimum amount of dark matter in
NGC 3198 inside 50 kpc.

the shape of the rotation curve for the halo component. The latter is fixed by the
observed rotation curve and the adopted disk. Defining the core radius of the halo
mass distribution with $\rho(R_{\mathrm{core}}) = 2^{-3/2}\rho(R = 0)$, we have $R_{\mathrm{core}} = (2^{3/2} - 1)^{1/\gamma}a$.
From this we find the — subjective — 95% confidence interval $9.6 < R_{\mathrm{core}} < 15.4$.
Thus, $R_{\mathrm{core}} = 12.5 \pm 1.5$ (1σ) kpc $= 4.7 \pm 0.6$ disk scale lengths. Note that this
is an upper limit for R_{core}: for smaller disk masses R_{core} decreases (see below).
Cumulative mass distributions for disk and halo are shown in Fig. 5. From this
figure it follows that the ratio of dark matter to visible matter inside the last point
of the rotation curve (at 30 kpc) is 3.9. The enclosed halo mass is 0.8 times the disk
mass at R_{25}; the enclosed halo mass is 1.5 times the disk mass at the Holmberg
radius. The total mass inside 30 kpc is $15 \times 10^{10}\ M_\odot$. Another property of interest
is the mass-to-light ratio of the disk; we find $M/L_B(\mathrm{disk}) \leq 3.6\ M_\odot/L_{B_\odot}$ and
$M/L_V(\mathrm{disk}) \leq 4.4\ M_\odot/L_{V_\odot}$.

The disk-halo model shown in Fig. 4 has the characteristic flat rotation curve
over a large part of the galaxy. Beyond 30 kpc it is a mere extrapolation, but the
observations inside 30 kpc do not show any sign of a decline, and the extrapolated
curve may well be close to the true one. To obtain an estimate of the minimum
amount of dark matter at large distances from the center we have also made a fit,
shown in Fig. 6, with a halo density law whose slope changes from -2 in the inner
region to -4 in the outer region:

$$\rho_{\text{halo}}(R) \propto \left[\left(\frac{a}{R_0} \right)^2 + \left(\frac{R}{R_0} \right)^2 + 0.08 \left(\frac{R}{R_0} \right)^4 \right]^{-1}, \qquad (2)$$

where $\rho_{\text{halo}}(R_0) = 0.0042 \, M_\odot \, \text{pc}^{-3}$, $a = 10 \, \text{kpc}$, and $R_0 = 8 \, \text{kpc}$; see BSS. Gradients for this density law are $d \, \log \rho / d \, \log R = -2, -3, -3.5$ at $R = 17, 32$, and 48 kpc, respectively. (These values, we feel, are not unreasonable. One could, of course, also simply truncate the halo at 30 kpc, but this is physically implausible.) The cumulative mass distribution for this case is shown in Fig. 5 as "minimum" halo. Using this curve, we find that the *minimum amount of dark matter associated with NGC 3198 inside 50 kpc is probably at least 6 times larger than the amount of visible matter*; thus, for the galaxy as a whole $M/L_B \geq 25 \, M_\odot / L_{B_\odot}$.

We now consider a family of disks with $V_{\text{max}} < 140 \, \text{km s}^{-1}$ (all with the same scale length); the previous results apply to the maximum disk case. We find that for each of the disks with $V_{\text{max}} < 140 \, \text{km s}^{-1}$ it is possible to find a matching halo described by Eq. (1), such that the sum of disk and halo fit the observed rotation curve. As an example, Fig. 7 shows a disk with mass equal to 30% of the maximum disk allowed. Even a fit with a halo only cannot be rejected straight away (see Fig. 8). The main difference between the resulting halos is the core radius: it decreases from $R_{\text{core}} = 12 \, \text{kpc}$ for the maximum disk mass case to 1.7 kpc when the dark halo dominates. The exponent γ of the halo density distribution is always close to 2.

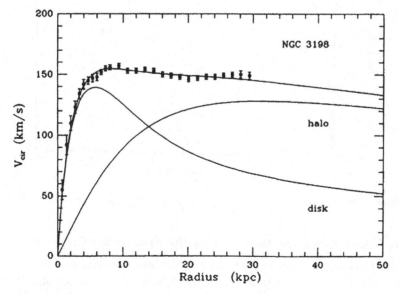

Fig. 6. Fit of "maximum" disk and "minimum" halo to observed rotation curve. The halo is represented by Eq. (2). See also Fig. 4.

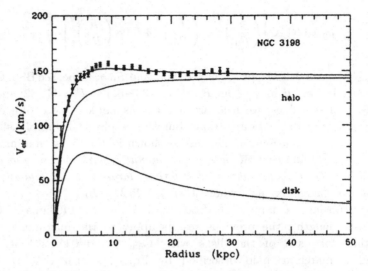

Fig. 7.   Fit of exponential disk with $M = 0.3 \times M_{\text{disk}}^{\text{max}}$ and halo to observed rotation curve. The parameters for the halo are $a = 1.3\,\text{kpc}, \gamma = 2.05, \rho(R_0) = 0.0063\ M_\odot\ \text{pc}^{-3}$.

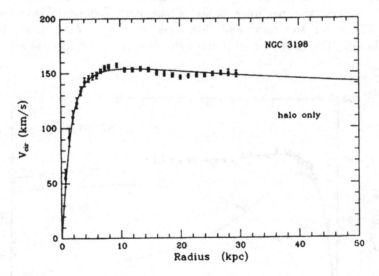

Fig. 8.   Fit of halo without disk; $a = 1.5\,\text{kpc}, \gamma = 2.25, \rho(R_0) = 0.0074\ M_\odot\ \text{pc}^{-3}$.

4.3. *An Upper Limit for the Mass of the Spheroidal Component*

Bulges of late-type spirals have fairly small effective radii (Simien and de Vaucouleurs 1983). Since there is also a rough relation between bulge luminosity and effective radius (Kormendy 1982), the effective radius of the bulge of NGC 3198 is expected to be quite small: $R_{\text{eff}} \approx 3\,\text{kpc}$ is a generous upper limit. We obtain an upper limit for the mass of a possible bulge component. The result is

$$M_{\mathrm{sph}}(30\,\mathrm{kpc})/M_{\mathrm{tot}}(30\,\mathrm{kpc}) < 0.026,$$

and

$$M_{\mathrm{sph}}(30\,\mathrm{kpc})/M_{\mathrm{disk}}^{\mathrm{max}}(30\,\mathrm{kpc}) < 0.12.$$

In reality the mass of the spheroidal component is probably much smaller, since in the optical data (Cheriguene 1975), which covers the region inside 2′ all the way to the center, there is no indication of a more rapid rise of the rotation in the innermost region.

Although the above upper limits are not very stringent, they indicate that our conclusions in the preceding paragraph are not affected by uncertainties regarding the properties of the spheroidal component.

5. Discussion

An important question left unanswered by the preceding analysis is the value of M/L for the disk. Should one seriously consider the case where the amount of visible matter is negligible with respect to the amount of dark matter (Fig. 8)? Or is the maximum disk case (Fig. 6) closer to the truth? There are three suggestive (but not definitive) reasons that support the latter possibility: (i) Measurement of mass and luminosity density in the solar neighborhood yields $M/L_V = 3.1 \pm 0.6\ M_\odot/L_{V_\odot}$ (Bahcall 1984). This value of M/L includes the dark material that must reside in the disk. The uncertainty represents an effective 95% confidence level. For NGC 3198 $M/L_V(\mathrm{disk}) \leq 4.4\ M_\odot/L_{V_\odot}$. (Note that M/L is proportional to the Hubble constant.) (ii) The shape of the rising part of the rotation curve agrees with that expected for a disk with scale length as given by the distribution of light. If the rotation curve were determined by the dark halo, such agreement would be a coincidence. (iii) The close relationship between luminosity of spiral galaxies and maximum circular velocity, implied by the small scatter in the Tully–Fisher relation, indicates that it is, after all, the amount of *visible* matter that determines the maximum rotation velocity in a galaxy. If this were not the case the amount of dark matter inside, say, 2.5 disk scale lengths must be related in a unique way to the amount of visible matter. (This may not be a strong argument; see Appendix.)

Of course, the overall flatness of rotation curves *also* implies a relation between the distributions of dark and visible matter. Indeed, indications at present are that rotation curves for spiral galaxies of all types *and* luminosities are approximately flat, or slightly rising, beyond the turnover radius of the disk (see Carignan 1983 and Carignan and Freeman, in preparation, for late-type spirals with M_B in the range -16 to -18). It is not clear yet whether this implies that the distributions of dark and visible matter are *closely* related. In analogy with the disk-bulge case, for which it is relatively easy to produce a flat rotation curve by combining a declining curve for the bulge with a rising one for the disk, it is also not difficult to make an approximately flat rotation curve beyond the turnover point of the disk by combining the declining curve for the disk with a rising one for the dark halo (see BSS). Carried one step further, this reasoning might be used as an argument against the motivation for Milgrom's (1983) proposal that Newtonian dynamics must be modified. Given the existence of dark halos, "fine tuning" of their properties to those of the visible matter would only be required if rotation curves of many galaxies turn out to be strictly flat until far beyond the turnover radius of the disk, like in NGC 3198.

So far we have assumed that dark halos are spherical. Alternatively, one might conjecture that the mass-to-light ratio of visible matter in disks of spiral galaxies increases with radius. In this case our calculations regarding the amount of dark matter require a downward adjustment by a factor of $\sim$1.5. It is too early to rule out this possibility, but a comparison of the thickness of H I layers with the vertical velocity dispersion of H I in face-on galaxies suggests that the dark matter needed to make the rotation curve flat cannot be hidden in a disk (van der Kruit 1981; Casertano 1983).

After these general remarks let us return to the case of NGC 3198. Having established that there is a large amount of dark matter, one would like to know its spatial distribution. Unfortunately, the present data provide only weak constraints. If we assume the mass of the disk to lie between 0.6 and 1 times the maximum value allowed by the rotation curve, we find that the core radius of the halo lies in the range 1.8–12.5 kpc. It is also of interest to compare the volume density of dark matter in the spherical halo with the density of H I. The surface density of H I outside $R = 6'$ can be represented by

$$\mu_{\mathrm{H\,I}}(R) \approx 120 e^{-R/h_{\mathrm{H\,I}}} M_\odot \ \mathrm{pc}^{-2},$$

where $h_{\mathrm{H\,I}} = 3'.3$. Consider a point in the midplane of H I at a distance of 27 kpc from the center, i.e. close to the outermost point on the rotation curve. Assume that the vertical density distribution of H I at this location is Gaussian with a 1σ scale height of 1 kpc. For the halo density law used for the maximum disk case (Fig. 4) we then find that the H I density *in the plane* still exceeds the density of dark matter by a factor of 8!

Finally, one may ask: how unique is NGC 3198? Can the results be generalized? The only unique feature that we know about for this galaxy is its relatively undisturbed, extended H I envelope, which allows an accurate determination of the rotation curve to large galactocentric distances. In other respects NGC 3198 is normal; in particular, its mass-to-light ratio inside one Holmberg radius is typical of Sc galaxies (Faber and Gallagher 1979). We thus have a solid determination of $M/L_B = 18 \ M_\odot/L_{B_\odot}$ inside 30 kpc, and a speculative estimate $M/L_B \geq 25$ inside 50 kpc, that are probably typical for galaxies of this morphology. The high M/L values found for binary galaxies (see the discussion by Faber and Gallagher) are therefore consistent with those for single galaxies over the same length scale.

Aknowledgements

We thank S. Casertano, K. C. Freeman, H. J. Rood, V. C. Rubin, and M. Schwarzschild for fruitful discussions and comments. T. S. vA. is grateful to the Institute for Advanced Study for its hospitality during the course of this work. This research was supported in part by NSF contract PHY-8440263 and PHY-8217352.

Appendix: Dark Matter and the Tully–Fisher Relation

Below we discuss, in a semiquantitative way, the effect of dark matter on the relation between luminosity and 21 cm profile width (T–F relation; Tully and Fisher 1977). We use the T–F relation in the following form:

$$L \propto V^{\alpha}, \qquad (A1)$$

where L is the total luminosity in a given wave band and V is the maximum circular velocity in the galaxy outside the bulge. Maximum circular velocity and profile width are, apart from a scale factor, nearly identical, as long as H I is present in the region around 2–3 disk scale lengths from the center. The value of α depends on the wave band and on the sample choice. For Sb–Sc galaxies, $\alpha \approx 3.5$ in B and $\alpha \approx 1.3$ in H (Aaronson and Mould 1983).

On the assumption that a galaxy consists of three components, a spheroid, an exponential disk, and a dark halo, Eq. (A1) can be rewritten as follows:

$$L \propto \{G[0.39M_{\text{disk}} + 0.45M_{\text{sph}}(2.2h) + 0.45M_{\text{halo}}(2.2h)]/h\}^{\alpha/2}, \qquad (A2)$$

where h is the disk scale length. Here we have made the additional assumptions that both spheroid and dark halo are spherical and that the location of maximum circular velocity coincides with the turnover point of the rotation curve of the disk. For reasonable choices of disk scale length and effective radius of the spheroid, most of the spheroid mass lies inside $2.2h$. Thus, for disk-dominated systems, we make a very small error — on the percent level — if we replace $0.45M_{\text{sph}}(2.2h)$ in Eq. (A2) by $0.39M_{\text{sph}}(\infty)$. [In fact, for 15 galaxies in Boroson (1981) with known h and R_{eff}, the median value of $0.45M_{\text{sph}}(2.2h)$ is $0.42M_{\text{sph}}(\infty)$.] Then

$$L \propto \{[M + 1.2M_{\text{halo}}(2.2h)]/h\}^{\alpha/2}; \qquad (A3)$$

M is the total visible mass of the galaxy, equal to $M_{\text{disk}} + M_{\text{sph}}$.

If the amount of dark matter inside 2.2 disk scale lengths is negligible, the T–F relation takes the simple form:

$$L \propto (M/h)^{\alpha/2}. \qquad (A4)$$

Such a relation is easy to understand if mass-to-light ratio and disk scale length vary with mass (see Burstein 1982). Adopting power-law relations, $M/L \propto M^a$ and $h \propto M^b$, one finds

$$\alpha = 2(1 - a)/(1 - b). \qquad (A5)$$

For a family of galaxies with disks with constant central surface density, $b \approx 0.5$ ($b = 0.5$ if there is no spheroidal component). Then $\alpha \approx 4(1 - a)$, and for M/L

independent of mass, $\alpha \approx 4$. This is the original explanation of Aaronson, Huchra, and Mould (1979) for the exponent 4 of the Tully–Fisher relation in the near-infrared.

Equation (A3) shows that the small observed spread in the T–F relation implies either a negligible amount of dark matter inside 2.2 disk scale lengths, or a correlation between the amounts of dark and visible matter. Such a correlation is indeed built-in: since disk scale length is known to depend on mass, the amount of dark matter inside 2.2 disk scale lengths also depends on mass. Thus, a relation between mass, luminosity, and disk scale length among galaxies as given in Eq. (A4) may not be destroyed by the presence of a moderate amount of dark matter, even if the density distribution of dark matter is largely independent of the visible matter.

References

Aaronson, M., Huchra, J. P., and Mould, J. R. (1979) *Astrophys. J.*, **229**, 1.

Aaronson, M. and Mould, J. (1983) *Astrophys. J.*, **265**, 1.

Bahcall, J. N. (1983) *Astrophys. J.*, **267**, 52.

———. (1984) *Astrophys. J.*, **287**, 926.

Bahcall, J. N., Schmidt, M., and Soneira, R. M. (1982) *Astrophys. J. (Letters)*, **258**, 123.

Bahcall, J. N. and Soneira, R. M. (1980) *Astrophys. J. Suppl.*, **44**, 73.

Begemann, F. (1985) in preparation.

Boroson, T. (1981) *Astrophys. J. Suppl.*, **46**, 177.

Bosma, A. (1981) *Astrophys. J.*, **86**, 1791.

Burstein, D. (1982) *Astrophys. J.*, **253**, 539.

Carignan, C. (1983) Ph.D. thesis, Australian National University.

Casertano, S. (1983) *Mon. Not. R. Astron. Soc.*, **203**, 735.

Cheriguene, M. F. (1975) in *Coll. Int'l. CNRS, La Dynamique des Galaxies Spirales*, ed. L. Weliachew, No. 241, p. 439.

de Vaucouleurs, G. (1959) in *Handbuch der Physik*, Vol. **53**, ed. S. Flügge (Berlin: Springer-Verlag), p. 311.

de Vaucouleurs, G., de Vaucouleurs, A., and Corwin, H. G. (1976) *Second Reference Catalogue of Bright Galaxies* (Austin: University of Texas Press) (RC2).

Faber, S. M. and Gallagher, J. S. (1979) *Ann. Rev. Astron. Astrophys.*, **17**, 135.

Freeman, K. C. (1970) *Astrophys. J.*, **160**, 811.

Holmberg, E. (1958) *Medd. Lund Astr. Obs.*, ser. 2. No. 136, p. 1.

Kalnajs, A. J. (1983) in *IAU Symposium 100, Internal Kinematics and Dynamics of Galaxies*, ed. E. Athanassoula (Dordrecht: Reidel), p. 87.

Kormendy, J. (1982) in *Morphology and Dynamics of Galaxies*, ed. L. Martinet and M. Mayor Saas-Fee (Sauverny: Geneva Observatory), p. 113.

Milgrom, M. (1983) *Astrophys. J.*, **270**, 365.

Rubin, V. C. (1983) in *IAU Symposium 100, Internal Kinematics and Dynamics of Galaxies*, ed. E. Athanassoula (Dordrecht: Reidel), p. 3.

Rubin, V. C., Ford, W. K. Thonnard, N., and Burstein, D. (1982) *Astrophys. J.*, **261**, 439.

Sandage, A. and Tammann, G. A. (1981) *A Revised Shapley-Ames Catalog of Bright Galaxies* (Carnegie Inst. of Washington Pub. No. 635).

Sanders. R. H. (1984) *Astron. Astrophys.*, **136**, L21.

Simien, F. and de Vaucouleurs, G. (1983) in *IAU Symposium 100, Internal Kinematics and Dynamics of Galaxies*, ed. E. Athanassoula (Dordrecht: Reidel), p. 375.

Tully, R. B. and Fisher, J. R. (1977) *Astron. Astrophys.*, **54**, 661.

van der Kruit, P. C. (1979) *Astron. Astrophyps. Suppl.*, **38**, 15.

———. (1981) *Astron. Astrophys.*, **99**, 298.

Wevers, B. M. H. R. (1984) Ph.D. thesis, Groningen University.

Young, P. J. (1976) *Astrophys. J.*, **81**, 807.

Chapter 3

SOME POSSIBLE REGULARITIES IN MISSING MASS*

John N. Bahcall and Stefano Casertano[†]

The Institute for Advanced Study,
Princeton, NJ, USA

The unseen matter in a sample of spiral galaxies exhibits simple regularities and characteristic numerical values.

1. Introduction

The missing mass (or light) problem has spawned many imaginative solutions involving neutrinos, gravitinos, axions, and other special proposals. Higher quality rotation curves now permit, for some galaxies, a quantitative measure of the distribution of dark matter inside galaxies. In this paper, we show first how remarkable is the observed simplicity of observed rotation curves, requiring fine-tuning if the visible and invisible matter are unrelated. We then draw attention to the fact that some *internal* properties of the dark mass seem to vary only over a relatively small range despite large differences in the *scale* properties of these galaxies (e.g. their masses and scale lengths). These characteristic numerical values in the galaxy data provide hints as to the nature of the missing matter and constraints on theories of galaxy formation. We discuss some of the implications for galaxy formation with *inos* and consider in somewhat more detail the possibility that the observed regularities may imply that the unseen matter is baryonic.

2. The Simplicity

Accurate rotation curves are now available for a number of galaxies, often extending well beyond the optical image of the galaxy. The rotation velocities do *not* decline in the outer reaches of galaxies where no appreciable light is visible. This lack of a decrease in rotation velocities is an example of the "missing light" or "missing mass" problem. For a few of these galaxies, it has been possible to estimate the different contributions of the visible and of the dark material to the total mass. The limited information already available is sufficient to hint at some surprising regularities, which may indicate the nature of the unseen material.

---

*Published in *Astrophys. J. Lett.* **293**, pp. L7–L10, 1985.
[†]Current address: Space Telescape Science Institute, Baltimore, MD 21218, USA.

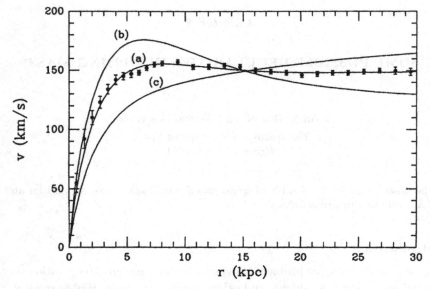

Fig. 1.   Observed rotation curve and mass models for NGC 3198. The disk scale length h is 2.7 kpc (Wevers 1984). The observed rotation curve (dots) is taken from van Albada et al. (1985), and extends almost to 30 kpc, or 11 disk scale lengths (2.4 Holmberg radii). There is no variation of the rotation velocity (within 10%) beyond two disk scale lengths. The mass models all consist of a disk with the same scale length as the observed light and a nearly isothermal halo with core radius of 7 kpc. The best fit (a) to the observed rotation curve is obtained with both the halo and disk masses equal to $M \equiv 3.1 \times 10^{10} M_\odot$ (inside the Holmberg radius). Models (b) and (c) are similar, except for a rescaling of masses in two components. Model (b) has 1.5 M in the disk and 0.5 M in the halo; model (c) has 0.5 M in the disk and 1.5 M in the halo.

The most striking feature of rotation curves is that there are no striking features. There is no overall change in the observed rotation curves that marks the transition between the inner region, in which the visible material dominates the gravitational field, and the so-called halo region which is filled with unseen material or missing mass. Collections of accurate optical and HI rotation curves (Bosma 1978; Rubin, Ford, and Thonnard 1980; Rubin et al. 1982) for many different galaxies exhibit this general characteristic of a featureless rotation curve. The most remarkable example is probably that of NGC 3198, an Sc galaxy whose 21 cm rotation curve extends well beyond two Holmberg radii. The rotation curve obtained by van Albada et al. (1985) (see dots in Fig. 1) remains constant *within 10%* over all the observed range between 7 kpc and 28 kpc. A few galaxies, like NGC 5907, do show some *local* feature (Casertano 1983a), but this is believed to be related to an abrupt truncation of the disk.

To illustrate the significance of the approximate constancy of the rotation curves, we show in Fig. 1 three different disk-halo mass models with the same geometrical properties. Model (a) fits the observations with an equal mass, M, in the disk and in the halo inside the Holmberg radius (van Albada et al. 1985). For model (b), the disk mass is 1.5 M, and the halo mass is 0.5 M. For model (c), the disk mass is reduced to 0.5 M, and the halo mass is increased to 1.5 M. Neither model (b) nor (c), with the disk and halo masses changed by only 50%, exhibit the same characteristic flatness as the observed rotation curve. Figure 1 shows that at least one parameter of a two-component mass model must be finely tuned in order to reproduce the observed flatness of the rotation curve.[a]

The simplest interpretation of Fig. 1 may be that there is only one type of galactic mass. We return to this point in the last section.

3. The Numerical Characteristics

For eight spiral galaxies detailed mass models exist and both visible and invisible components can be constrained. Table 1 summarizes the parameters of the mass distributions of these galaxies.

The masses of galaxies in our sample vary by almost a factor of one hundred. Though small, the sample covers a fairly wide range in galaxy types (from Sb to SBm) and external scale properties (mass, velocity, scale length). Despite these differences, the ratio M_H/M_D, of halo to disk mass inside the outer optical radius, seems to be *always* close to unity. Carignan and Freeman (1984) already noted this fact for the three late-type galaxies they studied.

The characteristic value of unity for M_H/M_D is not simply related to any obvious external property of galaxies. Until specific mass models were available, the ratios were not strongly constrained. There is no observational selection effect of which we are aware that can account for the relatively small variation that appears in Table 1.

In order to ensure homogeneity in the descriptions of the mass models, several choices had to be made, which will be detailed in the following. However, none of our conclusions is sensitive to the specific choices we have adopted: using somewhat different definitions we have found the same regularities.

The outer radius R_{out} has been defined as either the radius of the optical cutoff, if observed, or the radius at which the surface brightness, corrected for inclination and internal and galactic absorption, drops to 26.6 blue magnitudes per square arcsec (corresponding to $0.65\,L_\odot/pc^2$). We have derived the halo mass inside R_{out} directly from the mass models, and the disk masses from the published disk rotation curves. Sometimes small differences are present between the mass thus obtained and the value quoted by the original authors; but such differences are immaterial to our purpose. The disk surface density at R_{out} (used to evaluate the disk volume density ρ_D) is obtained by extrapolating an exponential disk, with the disk scale length h determined by the photometry. [This might increase the dispersion in ρ_H/ρ_D, since a pure exponential density distribution is not a good model for all

---

[a]If one varies two parameters, e.g. the core radius and the total mass of the halo within a fixed number of disk scale lengths, it is easy to construct a continuum of models with flat rotation curves. However, such models still occupy only a small fraction of the available parameter space.

J. N. Bahcall and S. Casertano

Table 1. Parameters of the Mass Distribution of Some Spiral Galaxies.

NGC	Type	h (kpc)	$R_{\rm out}$[a] (kpc)	$V_{\max}$ (km s^{-1})	$M_{\rm D}$ ($10^{10}M_\odot$)	$M_{\rm H}$ ($10^{10}M_\odot$)	$M_{\rm H}/M_{\rm D}$	z_0 (kpc)	$\rho_{\rm D}$ ($M_\odot$ pc^{-3})	$\rho_{\rm H}$ ($M_\odot$ pc^{-3})	$\rho_{\rm H}/\rho_{\rm D}$	References
MW	Sbc	3.5	14.7[b]	223	6.6[c]	9.5	1.4	0.50	0.0103	0.0047	0.46	1
891	Sb	4.9	21.0	231	9.1[d]	9.0	0.99	0.98	0.0042	0.00044	0.10	2
3198	Sc	2.68	12.7[e]	156	3.1	3.1	1.0	0.54[f]	0.0055	0.0014	0.26	3
4565	Sb	5.5	24.9	244	13.0[g]	14.6	1.1	0.79	0.0034	0.00097	0.28	4
5907	Sc	5.7	19.3	240	8.8	13.6	1.5	0.83	0.0102	0.0015	0.15	5
247	Sd	2.9	9.8[h]	100	1.20	1.08	0.90	0.58[f]	0.0067	0.0025	0.38	6
300	Sd	2.1	7.4[h]	83	0.63	0.54	0.84	0.42[f]	0.0080	0.0026	0.33	6
3109	SBm	1.6	5.8[h]	48	0.16	0.15	0.93	0.32[f]	0.0041	0.0016	0.38	6

[a] Inside $R_{\rm out}$, and excluding the central peak (if present).
[b] Assumed ($= 4.2h$) from central surface brightness.
[c] Includes spheroid and nuclear component; 5.3 for pure disk.
[d] Includes spheroid; 8.7 for pure disk.
[e] Assumed ($= 4.75h$) from central surface brightness.
[f] Assumed from $z_0 \approx 0.2h$.
[g] Includes spheroid; 9.0 for pure disk.
[h] From photometry, corrected for inclination, etc.
References: (1) Bahcall, Schmidt, and Soneira 1982. (2) Bahcall 1983. (3) van Albada et al. 1985. (4) Casertano 1983b. (5) Casertano 1983a. (6) Carignan and Freeman 1985.

galaxies; see Carignan (1984a, b), and van der Kruit & Searle (1981a, b).] The thickness is expressed in terms of z_0, the parameter appearing in the sech^2 law; we recall that, in first approximation, z_0 is twice the exponential scale height of the disk.[b] For the four galaxies in the sample for which the thickness cannot be measured, we have used the rule $z_0 \approx 0.2h$, which is approximately valid for the other galaxies.

The following equations summarize the relatively well determined parameters that characterize the internal properties in all the mass models in our sample; they are evaluated at R_{out}:

$$M_{\mathrm{H}}/M_{\mathrm{D}} \approx 1, \tag{1}$$

$$\rho_{\mathrm{H}}/\rho_{\mathrm{D}} \approx 0.3, \tag{2}$$

$$\rho_{\mathrm{H}} \approx 0.0015\, \mathrm{M}_{\odot}/\mathrm{pc}^3. \tag{3}$$

These numerical values constitute benchmarks relative to which theories of spiral galaxy formation can be tested.

Relations (1) and (2) are satisfied with a total dispersion of only ±30% in the values of $M_{\mathrm{H}}/M_{\mathrm{D}}$ and ±50% for $\rho_{\mathrm{H}}/\rho_{\mathrm{D}}$. The tightness of relation (1) is especially striking since the disk and halo masses separately vary by a factor of 100. The third relation shows much more scatter, but it may be important because it sets the dimensional scale for the relevant physical processes.

A relation similar to Eq. (1) can be obtained by assuming a Tully–Fisher relation that is independent of Hubble type, a constant central surface brightness, and a universal ratio of mass to luminosity. However, the actual central brightnesses vary in our sample by about a factor of 7, and the mass–luminosity ratios differ for the individual galaxies by a factor of order 3 (according to the individual estimates of the original authors). From the previously known relations, we would have expected a total dispersion larger than the observed 30%.

One might worry that relation (1) is an artifact of the method used in deriving the mass models. In fact, if the rotation curves were always flat and the disk was always assumed to dominate the rotation curve in the inner parts (the "maximum disk" assumption), one would expect a constant ratio of disk to halo mass inside a specified number of disk scale lengths. However, the mass models have been derived by the different authors using a variety of methods. Only in four cases (NGC 3198, van Albada et al. 1985, and NGC 247, 300 and 3109, Carignan & Freeman 1984) has the "maximum disk" method been used. For the Milky Way, the disk mass has been normalized (Bahcall, Schmidt, & Soneira 1982) to the value in the solar neighborhood; for NGC 5907 the disk is "measured" through the amplitude of the

---

[b]The sech^2 law, $\rho(z) = \rho_0 \cdot \mathrm{sech}^2(z/z_0)$, is probably not a good model of the vertical density distribution at the outer radius of the disk (Bahcall and Casertano 1984). However, z_0 is an adequate measure of the thickness of the disk.

truncation blip (Casertano 1983a); and for the remaining two galaxies, NGC 891 and NGC 4565, the authors (Bahcall 1983 and Casertano 1983b, respectively) have used a standard mass-to-light ratio — compatible with the rotation curve — to obtain the disk mass. Given this variety of methods, it is surprising that relation (1) is valid to 30%.

A word of caution is in order. The galaxies we consider are among the best studied and the most constrained in their mass distribution. But, even for these cases, the mass models are far from unique. For example, there is no *dynamical* evidence (except for the Milky Way and perhaps for NGC 5907) that the disk has any appreciable mass. In all but these two cases, it would be possible to make models in which the mass of the disk is negligible and the halo dominates the gravitational field. The numbers in Table 1, although state-of-the-art, require plausibility arguments (see preceding paragraph) in addition to the observed rotation curves.

The only criterion used in selecting the galaxies in our sample is the existence of a detailed mass model that reproduces high-quality observations. Therefore, the galaxies included do *not* constitute a well-defined population in any sense. In particular, there are no galaxies earlier than Sb and no normal barred spirals. A number of additional galaxies must be observed carefully and analyzed in detail in order to test whether or not the relations that we have summarized here are general. In our discussion (Sec. 5), we will assume that the regularities we have found are indeed characteristic of a large class of spiral galaxies.

4. The Local Missing Mass

In at least one well studied case, the missing matter cannot *all* be in a dissipationless component. Bahcall (1984a, b) has shown that about half the matter in the solar neighborhood is unseen and has an exponential scale height that is *less than* 0.7 kpc (twice the scale height of the old disk stars). This result is obtained by a modern reanalysis of the Oort limit using three stellar samples with improved data and theoretical computations, combined with the upper limit on the total mass within the solar position that is inferred from the galactic circular velocity at the sun. Therefore, at least some of the dark matter is in a dissipational component, which has been able to collapse into a highly flattened disk.

5. Implications

What is the meaning of the regularities described in Secs. 2 and 3? One possibility, considered by Milgrom (1983), is that the law of gravitation differs from the Newtonian theory at the very small accelerations that are characteristic of the outer reaches of galaxies. In fact, Milgrom's modification to the law of gravitation would explain both the observed flatness of the rotation curves and the constancy of M_H/M_D (in Milgrom's theory the "halo" is not real). We will not discuss this alternative. We prefer to consider instead the alternatives that do not require changes in the accepted laws of physics.

In more popular scenarios, many authors have suggested that the unseen material is intrinsically different from the visible matter: the former being made of exotic particles and the latter being made of ordinary stuff, like stars and gas. Thus, the observed material would be dissipational and the unseen material dissipationless.

The phenomenological relations that we have discussed between visible and invisible matter can be summarized as follows:

(i) Featureless rotation curves require fine tuning if the inner galactic material (stars and gas) is physically different from the halo (dark mass).

(ii) There are regularities in the mass models [Eqs. (1)–(3)] that suggest a close relation between the visible and invisible matter.

(iii) The missing matter in the disk is dissipational, like the visible material.

The regularities listed above must be telling us something important about galaxy formation if there is indeed a fundamental distinction between visible and invisible matter. The significance of the regularities may be revealed by detailed calculations (or numerical simulations) of the formation of galaxies from primordial perturbations in the early universe. One possible mechanism that might provide a natural explanation of some of the observed regularities, based on the reaction of the halo to the gravitational field of the collapsing disk, is currently being considered (J. Barnes, private communication).

On the basis of the above-cited phenomenological relations between visible and invisible matter, we now explore the possibility that both the invisible and the visible matter are baryonic. In this picture, some of the original mass condensed into a disk and formed visible stars, changing only slightly the overall mass distribution and the rotation curve. This hypothesis accounts in a plausible way for featureless rotation curves (see Fig. 1). The difference between seen and unseen material is attributed to the ability of galactic matter to form easily observable stars — stars massive enough to burn hydrogen ($> 0.085\,M_\odot$) — above the critical mass density represented by Eq. (3). Regularities (1) and (2) can then be interpreted as representing the efficiency of star formation.

Where could this dark baryonic mass be located? Several authors have stated that it is "unlikely" that the unseen mass is in the disk (Casertano 1983a; Lake 1981; van der Kruit 1981), but the observational evidence is not conclusive. It is at least conceivable that the dark material is all in a thick disk, a scenario that can be tested by more realistic theoretical studies of the stability of galactic disks. A less extreme possibility is that part of the unseen matter has collapsed to a disk and some invisible mass is still in a rounder component. However, if the baryonic picture is to account in a natural way for the regularities expressed in Eqs. (1)–(3), the dark mass must have collapsed to much the same extent as has the disk.

The unseen material is probably not in a gaseous form (see, e.g. Bergeron and Gunn 1977). The existing observational upper limits on H I emission at large galactocentric distances show that the halo contains only a relatively small amount of neutral gas. Similarly, a hot gaseous halo would produce — by bremsstrahlung — a much higher total X-ray emission ($>10^{42}$ erg sec^{-1}) than is observed for typical spiral galaxies (Fabbiano, Trinchieri, and Macdonald 1984). At an intermediate temperature like 10^4 K, the emission measures in the models discussed here are sufficiently large that they could be observed with existing techniques. The best cases for detecting intermediate temperature gas are edge-on spirals for which the emission measure would be expected to be $\sim$40 cm^{-6} pc outside the Holmberg radius on the major axis and could be a factor of several higher along the minor axis. However, the emission measure for the Milky Way would be $\sim$300 cm^{-6} pc, more than two orders of magnitude larger than the most recent measurements permit (Reynolds 1984). At least for the Galaxy, the unseen material cannot be in gas at about 10^4 K.

The only remaining baryonic candidates for the unseen material are: (1) massive collapsed objects, or (2) stellar-like objects that are not massive enough to burn hydrogen. None of the existing observational constraints are in conflict with these suggestions. Either possibility requires that stars form in the non-collapsed phase with an initial mass function quite different from that of the Population I disk stars. In addition, the formation of stars from low density primordial gas must have blown away nearly all of the gas by a combination of winds, shocks and radiation pressure, or the gas must have been removed by interaction with the intergalactic medium (see Bergeron and Gunn 1977). Elliptical galaxies have managed to largely rid themselves of gas; perhaps the baryonic halos of spiral have been similarly successful.

Acknowledgements

We are grateful to many colleagues, and especially to T. van Albada, J. Barnes, D. Gilden, P. Goldreich, C. McKee, J. Oort, S. Phinney, V. Rubin, M. Schmidt, S. Tremaine, and S. White, for discussions of the missing matter problem. This work was supported in part by NSF grant PHY-8440263.

References

van Albada, T. S., Bahcall, J. N., Begeman, K. & Sancisi, R. (1985) Preprint.
Bahcall, J. N. (1983) *Astrophys. J.*, **267**, 52.
Bahcall, J. N. (1984a) *Astrophys. J.*, **276**, 169.
Bahcall, J. N. (1984b) *Astrophys. J.*, **287**, 926.
Bahcall, J. N. & Casertano, S. (1984) *Astrophys. J. Lett.*, **284**, L35.
Bahcall, J. N., Schmidt, M. A. & Soneira, R. M. (1982) *Astrophys. J. Lett.*, **258**, L23.
Bergeron, J. & Gunn, J. E. (1977) *Astrophys. J.*, **217**, 892.
Bosma, A. (1978) Ph.D. thesis, Groningen.
Carignan, C. & Freeman, K.C. (1984) Preprint.
Carignan, C. (1984a) Preprint.
Carignan, C. (1984b) Preprint.
Casertano, S. (1983a) *Mon. Not. R. Astron. Soc.*, **203**, 735.
Casertano, S. (1983b) *Thesis*, Pisa.

Fabbiano, G., Trinchieri, G. & Macdonald, A. (1984) *Astrophys. J.*, **284**, 65.

Lake, G. (1981) in *Cosmology and Particles*, ed. J. Audouze, P. Crane, T. Gaisser, D. Hegyi & J. Tran Tranh Van (Frontiers: Dreux), p. 331.

Milgrom, M. (1983) *Astrophys. J.*, **270**, 365.

Reynolds, R. J. (1984) Preprint.

Rubin, V. C., Ford, W. K. & Thonnard, N. (1980) *Astrophys. J.*, **238**, 471.

Rubin, V. C., Ford, W. K., Thonnard, N. & Burstein, D. (1982) *Astrophys. J.*, **261**, 439.

van der Kruit, P. C. (1981) *Astron. Astrophys.*, **99**, 298.

van der Kruit, P. C. & Searle, L. (1981a) *Astron. Astrophys.*, **95**, 105.

van der Kruit, P. C. & Searle, L. (1981b) *Astron. Astrophys.*, **95**, 116.

Wevers, B. M. H. R. (1984) Ph.D. thesis, Groningen.

Chapter 4

EVOLUTION OF GLOBULAR CLUSTERS AND THE GLOBULAR CLUSTER SYSTEM – I

J. P. Ostriker

Princeton University Observatory
Princeton, NJ 08544, USA

C. Thompson*

Joseph Henry Laboratories
Princeton, NJ 08544, USA

In this first part of these lectures, we review the observed properties and dynamical evolution of globular clusters and the globular cluster system. We concentrate on the behavior of a cluster during core collapse, and discuss how the formation of binaries in the core may heat the cluster and reverse the collapse. We review the various processes which disrupt clusters, and their cumulative effect on the globular cluster system. We briefly consider the constraints which may be placed on baryonic dark matter in the Galactic halo.

Later in the second part, we review the properties of positive energy perturbations in an expanding universe. We construct self-similar solutions of various types for blast waves in both static and expanding media. We concentrate on those features which distinguish blast waves in these two types of media. We briefly consider the stability of the thin shells formed by cosmological blasts, and what happens when shells interact.

1. Globular Clusters

Globular clusters are one of the oldest subjects of theoretical study in astronomy; only the dynamics of the solar system, and binary star systems, have been investigated in greater detail. A globular cluster, more so even than a galaxy, is the prototypical self-gravitating system of many point masses. The equilibrium structure of globular clusters has been adequately understood for some time, but only recently has much progress been made in unravelling their evolution.

At first sight, the study of globular clusters may seem to reveal little about dark matter; for example, there is no evidence that globular clusters contain dark matter (but, see Peebles 1984). We will find, nonetheless, that globular clusters do place some indirect constraints on the distribution and composition of dark matter in the Galaxy.

---

*Current address: CITA, University of Toronto, Toronto, ON M5S 3H8, Canada.

Detailed compendia of recent work on globular clusters can be found in the proceedings of IAU Symposia 113 and 126. For an expository account, the reader is referred to the forthcoming book by Spitzer (1987).

2. Basic Properties of Globular Clusters

A globular cluster is a nearly spherical assembly of stars, bound by its own gravity. About 150–200 clusters are distributed through the halo of the Galaxy, with approximate spherical symmetry about its center. The average mass of a cluster is $\simeq 10^5 M_\odot$, and the mass to light ratio is typically $M/L_V \simeq 1 - 2$, in solar units. The space density of clusters is strongly concentrated toward the Galactic center, falling off with distance R from the center as R^{-3} for $R < 10\,\mathrm{kpc}$, and as R^{-4} for $R > 10\,\mathrm{kpc}$. Several clusters lie beyond $50\,\mathrm{kpc}$. The core radius of the cluster distribution is approximately $1\,\mathrm{kpc}$. We cannot determine directly the number of clusters within $1\,\mathrm{kpc}$; but a study of the decay of cluster orbits toward the Galactic center, indicates that this number may be about 50 (Oort 1977).

In an individual cluster, one may distinguish three characteristic distances, as measured from the center. The *core radius* R_c is defined to be the radius at which the surface brightness is half its central value, and takes on the values 0.3 to 3 pc. The *"tidal" radius* R_t is the radius at which the extrapolated surface brightness reaches zero. One finds $R_t/R_c \sim 10$–100. The radius R_h containing half the projected light, and presumably the mass, of the cluster is found empirically to satisfy the approximate relation $R_h \sim \sqrt{R_t R_c}$. The velocity dispersion of the stars parallel to the line of sight is $V_\parallel = 3$–$5\,\mathrm{km\,s^{-1}}$. From the virial theorem we obtain

$$2 \cdot \frac{3}{2} M_h V_\parallel^2 \simeq \left(G M_h^2 / 2 R_h \right), \quad \text{or} \quad M_h \simeq 6 R_h V_\parallel^2 / G = 10^5 M_\odot$$

for typical values $V_\parallel = 5\,\mathrm{km\,s^{-1}}$ and $R_h = 5\,\mathrm{kpc}$.

The measured distribution of cluster velocities in the Galactic halo is consistent with an isotropic distribution (Frenk and White 1980), although there is evidence that some clusters rotate at $< 50\,\mathrm{km\,s^{-1}}$ (Zinn 1985). There is no evidence of any trend in these rotational velocities, if they exist, with radius. (But see the contribution of Tremaine and Lee on the dynamics of satellite dwarfs.)

There is an apparent division (Zinn 1985) in the properties of globular clusters between those within $\sim 5\,\mathrm{kpc}$ of the Galactic center (the "Disk" population) and those outside this radius (the "Bulge" population). The Disk population appears to be flattened (with an axis ratio of 2:1), rotating, and metal rich ($[Fe/H] > -0.8$). The Bulge population is spherically symmetric and relatively metal poor ($-2 < [Fe/H] < -0.8$). There is a clear negative gradient in the mean density and core density ρ_c of Bulge clusters, outward from the Galactic center.

The age of globular clusters is determined by comparing the position of the observed main sequence turn-off on the HR diagram, to that inferred from the observed helium abundance and metallicity of the cluster. The main uncertainties arise from the calibration of the distance scale to the clusters (using RR Lyrae stars, or comparing the main sequence luminosity with that of a nearby cluster), and the indirect determination of helium abundance. The ages thus determined typically lie in the range $(12-17) \times 10^9 \, \mathrm{yr}$. Recent work has suggested that much of this spread may be real, and that metal rich clusters are younger.

The stellar population of globular clusters is dominated by dwarfs, with red giants comprising 4%, degenerate dwarfs anywhere up to 30%, and neutron stars perhaps a few percent of the mass. In all, globular clusters contribute a fraction $\sim$1% of the optical luminosity of the spheroidal component of the Galaxy.

A half-dozen strong X-ray sources, with luminosities 10^{35}–$10^{37} \, \mathrm{erg\, s^{-1}}$, have been found in globular clusters. These are interpreted as neutron star–main sequence binaries. A much larger number of weaker X-ray sources, with luminosities $10^{33} \, \mathrm{erg\, s^{-1}}$, have also been detected. These are probably degenerate dwarf–main sequence binaries. The relative concentration of X-ray binaries in globular clusters indicates that binaries may form by two-body tidal capture (Fabian, Pringle and Rees 1975) in the dense cores, where the density of stars is in the range 10^3–$10^5 \, \mathrm{pc^{-3}}$.

3. Equilibrium Structure of Globular Clusters

Let us now consider the equilibrium configuration of a spherical star cluster, bound by its own gravity. First, we compare the time for an individual star to complete one orbit, with the time for that orbit to be modified by interactions with other stars. Even though the density of stars in the core of a cluster is large, individual orbits change only slowly. Thus, the idea of an equilibrium configuration is well-defined.

The dynamical time of passage of a star with velocity V_* through the cluster is

$$t_{\mathrm{d}} \sim \frac{R_{\mathrm{h}}}{V_*} = 10^6 \left(\frac{R_{\mathrm{h}}}{5\,\mathrm{pc}}\right) \left(\frac{V_*}{5\,\mathrm{km\,s^{-1}}}\right)^{-1} \mathrm{yr}. \tag{1}$$

We have substituted a typical line-of-sight velocity for V_*. Now, a typical encounter between stars occurs at a separation $\Delta r \sim N_*^{-1/3} R_{\mathrm{h}} \sim 0.1 \, \mathrm{pc}$ where $N_* \sim 10^5$ is the number of stars in the cluster. The characteristic deflection of each star in one encounter is small:

$$\frac{\Delta V_\perp}{V_*} \sim \frac{G m_*}{(\Delta r)^2} \frac{\Delta r}{V_*^2} \sim \frac{1}{N_*} \frac{R_{\mathrm{h}}}{\Delta r} \sim N_*^{-2/3}. \tag{2}$$

Now let us sum the effects of many encounters. The component of a star's velocity, perpendicular to its initial direction of motion, grows diffusively as

$$\frac{\partial}{\partial t} \left(\frac{V_\perp}{V_*}\right)^2 \simeq \int_{\Delta r(\min)}^{\Delta r(\max)} 2\pi \Delta r \, d(\Delta r) n_* V_* \left(\frac{\Delta V_\perp}{V_*}\right)^2$$

$$= 2\pi n_* \frac{G^2 m_*^2}{V_*^3} \ln\left(\frac{R_{\mathrm{h}}}{G m_*/V_*^2}\right). \tag{3}$$

The cumulative effect of distant encounters diverges with Δr, so we impose a cut-off at the size R_{h} of the cluster. In addition, the deflection is no longer small at separations less than Gm_*/V_*^2, and we choose this as the lower cut-off. Our main conclusion is that the deflection rate due to distant encounters exceeds that due to near encounters (at $\Delta r < Gm_*/V_*^2$) by a factor $\ln \Lambda = \ln(0.4N_*)$. The coefficient in the logarithm and the overall numerical coefficient are given by a more careful treatment (cf. Spitzer 1987) which we will summarize shortly.

We can now compute the time for a single orbit to evolve significantly — the *relaxation time*, t_r. We find that it greatly exceeds the dynamical time, and in many cases is comparable to the age of the cluster:

$$t_r = \left[\frac{\partial}{\partial t} \left(\frac{V_\perp}{V_*} \right)^2 \right]^{-1} = \frac{N_*}{\beta \ln(0.4\,N_*)} t_d$$

$$\sim \frac{1}{\beta} 10^4 \, t_d \sim \frac{1}{\beta} 10^{10} \, \text{yr}, \quad (N_* = 10^5). \qquad (4)$$

This represents the relaxation time at the half-mass radius. Our estimate (3) indicates that $\beta \sim 10$. In the core of a cluster, where the density of stars is high, the relaxation time is much shorter. As we will see, core evolution is a central topic in contemporary research on globular clusters.

To construct a self-consistent model of a globular cluster, one may simply choose the distribution function f. Recall that the number of stars in an infinitesimal volume of phase space is

$$dN = f(\mathbf{r}, \mathbf{v}, t) d^3r \, d^3v. \qquad (5)$$

The fact that $t_r \gg t_d$ allows us to neglect individual two-body encounters in calculating the equilibrium configuration of the cluster. Liouville's theorem then tells us that the distribution function evolves according to

$$\frac{Df}{Dt} = \frac{\partial f}{\partial t} - \nabla \phi \cdot \frac{\partial f}{\partial \mathbf{v}} + \mathbf{v} \cdot \frac{\partial f}{\partial \mathbf{r}} = 0. \qquad (6)$$

This is the *collisionless Boltzmann equation*, or the *Vlasov equation*. We have substituted $d\mathbf{v}/dt = -\nabla \phi(\mathbf{r}, t)$, where ϕ is the coarse-grained gravitational potential at $\mathbf{r}$ due to all the stars in the cluster. That is, in neglecting two-body encounters we are using $\phi = \tilde{\phi}$, where

$$\nabla^2 \tilde{\phi} = -4\pi G \tilde{\rho}. \qquad (7)$$

Here, $\tilde{\rho} \equiv (\Delta V)^{-1} \int_{\Delta V} \rho \, dV$ is the stellar mass density averaged over a volume ΔV which contains many stars, but is small compared to the size of the cluster.

The number density of stars is related to the distribution function by $n_* = \int f(\mathbf{r}, \mathbf{v}, t) d^3v$. The local mean velocity $\mathbf{u}(\mathbf{r}) = n_*^{-1} \int \mathbf{v} f d^3v$, and the velocity dispersion $V_m^2(\mathbf{r}) = n_*^{-1} \int (\mathbf{v} - \mathbf{u})^2 f \, d^3v$.

When a globular cluster has attained an equilibrium configuration, the distribution function f is independent of time, $\partial f/\partial t = 0$. For each orbit, it is easy to find one or more conserved integrals of motion: namely, the energy per unit mass $E = \frac{1}{2}v^2 + \phi(r)$; and also, the component J_z of the specific angular momentum, if the cluster is axially symmetric; or all three components $\mathbf{J} = \mathbf{r} \times \mathbf{v}$ of the angular momentum, if the cluster is spherically symmetric.

Now, an immediate consequence of Liouville's theorem (6) is that the distribution function depends only on the integrals of motion. This is sometimes known as Jeans' theorem in the literature on stellar dynamics. For an orbit in three dimensions, there are generally *six* independent integrals of motion (five when $\partial f / \partial t = 0$), and the integrals displayed above do not exhaust the list.

We conclude that the distribution function in a spherical system may be chosen to be $f = f(E, J^2)$; but this is not the most general form that f may take. Now, in an isolated system, E and J^2 may each take on only a limited range of values. Namely, if a star is bound to the cluster, then $E < 0$; and, given a value of E, the angular momentum cannot be greater than that of a circular orbit.

The two most common choices for the distribution function are as follows. In a "polytrope" model, one takes

$$f \propto (-E)^p \tag{9}$$

which implies $\rho \propto \phi^{p+3/2}$. The choice $p = -\frac{3}{2}$ corresponds to a uniform sphere, and $p = \frac{7}{2}$ is the *Plummer model*, which is known in the literature of stellar structure as a polytrope of index $n = 5$. In this latter case, the density and potential are

$$\rho(R) = \frac{3M}{4\pi R_0^3} \frac{1}{(1 + R^2/R_0^2)^{5/2}},$$
$$\phi(R) = \frac{GM}{R_0} \frac{1}{(1 + R^2/R_0^2)^{1/2}} = -2V_{\mathrm{m}}^2(R). \tag{10}$$

The *King model* (1966) has an isothermal distribution function with an upper cut-off in energy (representing the outer radius of the cluster),

$$f = \begin{cases} K(e^{-\beta E} - e^{-\beta E_0}) & E < E_0 \\ 0 & E > E_0 \end{cases}. \tag{11}$$

The "temperature" β^{-1} of the stars is related to the central velocity dispersion in the cluster by

$$\beta^{-1} = \frac{1}{3} V_{\mathrm{m}}^2(0) = V_{\|}^2(0).$$

Here, $V_{\|}^2$ is the line-of-sight velocity dispersion. For example, if the tidal field of the Galaxy determines the outer radius of the cluster, then $(2R_{\mathrm{t}}/R_{\mathrm{g}}) \cdot GM_{\mathrm{g}}/R_{\mathrm{g}}^2 = GM_{\mathrm{cl}}/R_{\mathrm{t}}^2$ and the tidal radius is

$$R_{\mathrm{t}} = \left(\frac{M_{\mathrm{cl}}}{2M_{\mathrm{g}}} \right)^{1/3} R_{\mathrm{g}}, \tag{12}$$

where M_{cl} is the cluster mass, R_{g} the distance from the center of the Galaxy, and M_{g} the mass of the Galaxy interior to R_{g}. That is, we choose for the energy cut-off

$$E_0 = \phi(R_{\mathrm{t}}) = -\frac{GM_{\mathrm{cl}}}{R_{\mathrm{t}}}. \tag{13}$$

Note also that the core radius R_c is defined in terms of the central velocity dispersion and central density by

$$R_c = \left(\frac{3V_m^2(0)}{4\pi G \rho(0)} \right)^{1/2}. \tag{14}$$

In summary, a King model may be parametrized by the dimensionless ratio R_c/R_t.

We may relate the escape velocity from the cluster to the virial velocity as follows. The mean escape velocity is

$$\frac{1}{2} M_{cl} \langle V_{esc}^2 \rangle = \frac{1}{2} \int dr 4\pi r^2 \rho_*(r) V_{esc}^2(r) = - \int dr 4\pi r^2 \rho_*(r) \phi(r) = -2E_\phi,$$

where E_ϕ is the gravitational binding energy of the cluster and M_{cl} is its mass. By the virial theorem, $E_\phi = -2 \cdot \frac{1}{2} M_{cl} \langle V_m^2 \rangle$ so we have the following general relation,

$$\langle V_{esc}^2 \rangle = 4 \langle V_m^2 \rangle. \tag{15}$$

The mean escape velocity is only a factor of two larger than the mean r.m.s. velocity of stars in the cluster. This is our first clue that the equilibrium configurations under consideration may not persist over many relaxation times.

Indeed, it is clear that stars on the exponential tail of the velocity distribution will *evaporate* from the cluster, reducing its mass but increasing its binding energy (Ambartsumian 1938; Spitzer 1940). In spite of this, the stars outside the core *expand*. The angular momentum of an individual star is conserved during the gradual loss of mass. For example, if it is in a circular orbit at radius R, then $j^2 = GM(<R)R$ is conserved, and

$$\frac{1}{R} \frac{\partial R}{\partial t} = - \frac{1}{M(<R)} \frac{\partial M(<R)}{\partial t}. \tag{16}$$

In order to avoid a contradiction, we conclude that the density and velocity dispersion of stars *in the core* must increase as the cluster loses mass. The evaporation of stars from the cluster leads eventually to core *"collapse"*.

So far, we have only considered the effect of successive distant encounters on the component of a star's velocity *perpendicular* to its initial direction of motion. We found that the growth of $V_\perp$ was diffusive, because the expected deflection in one encounter $\langle \Delta V_\perp \rangle = 0$. Once again, let us consider the motion of a test star with velocity V_* and mass m_* through a background field of stars of mean density ρ_f and mass m_f. During distant encounters, the field stars absorb momentum and energy from the test star. It follows that the expected change in the *parallel* component of the test star's velocity during a distant encounter, does not vanish.

We may estimate the deceleration of the star, by *dynamical friction*, as follows. The original treatment is due to Chandrasekhar (1942). Suppose for simplicity that the field stars are at rest. The velocity $\Delta V_\parallel$ imparted to a field star during one encounter at impact parameter Δr is $\Delta V_\parallel \sim \Delta V_\perp^2/V_*$. The dynamical friction time is therefore

$$t_{\mathrm{df}}^{-1} = \frac{1}{m_* V_*} \frac{d(m_* V_*)}{dt} = \int d(\Delta r) 2\pi \Delta r \cdot \rho V_* \Delta V_{\parallel} \simeq 2\pi \frac{G^2 \rho m_*}{V_*^3} \ln \Lambda, \quad (17)$$

where $\ln \Lambda = \ln(0.4 N_*)$ as before. Assuming that $V_*^2 \sim V_{\mathrm{m}}^2$, we may compare t_{df} with the two-body relaxation time:

$$t_{\mathrm{df}} \sim t_{\mathrm{r}} \cdot \left(\frac{m_{\mathrm{f}}}{m_*}\right). \quad (18)$$

A massive star slows down before its orbit is significantly deflected.

Spitzer (1987) provides more exact results. These are

$$\left\langle \frac{\partial}{\partial t} \Delta V_{\parallel} \right\rangle = -\frac{12\pi G^2 \rho_{\mathrm{f}} (m_{\mathrm{f}} + m_*)}{V_{\mathrm{m,f}}^2} \ln \Lambda \cdot G(x), \quad (19)$$

$$\left\langle \frac{\partial}{\partial t} \Delta V_{\parallel}^2 \right\rangle = \frac{12\pi G^2 \rho_{\mathrm{f}} m_{\mathrm{f}}}{V_{\mathrm{m,f}}^2} \ln \left(\Lambda \cdot \frac{G(x)}{x}\right), \quad (20)$$

and

$$\left\langle \frac{\partial}{\partial t} \Delta V_{\perp}^2 \right\rangle = \frac{12\pi G^2 \rho_{\mathrm{f}} m_{\mathrm{f}}}{V_{\mathrm{m,f}}^2} \ln \left(\Lambda \cdot \frac{\Phi(x) - G(x)}{x}\right), \quad (21)$$

where $V_{\mathrm{m,f}}^2$ is the velocity dispersion of the field stars,

$$x = \sqrt{3 V_*^2 / 2 V_{\mathrm{m,f}}^2},$$

$$\Phi(x) \equiv \frac{2}{\sqrt{\pi}} \int_0^x e^{-y^2} dy,$$

$$G(x) \equiv \frac{1}{2} x^{-2} [\Phi(x) - x\Phi'(x)].$$

Heavy stars, of mass m_*, lose energy to lighter stars, of mass m_{f} in a time

$$t_{\mathrm{eq}} = \frac{(V_{\mathrm{m,f}}^2 + V_{\mathrm{m,*}}^2)^{3/2}}{8\sqrt{6\pi} G^2 m_* \rho_{\mathrm{f}} \ln \Lambda}, \quad (22)$$

in the limit where the kinetic energy of the heavy stars greatly exceeds that of the lighter stars. Indeed, the initial velocity dispersion of the heavy stars in a cluster (after violent relaxation; see below) is expected to be comparable to that of the lighter stars, $V_{\mathrm{m,*}}^2 \sim V_{\mathrm{m,f}}^2$.

Applying these results to a globular cluster composed of stars of constant mass m_*, one finds that the relaxation time at the half-mass radius is (Spitzer and Hart 1971),

$$t_{\mathrm{rh}} = \frac{V_{\mathrm{h}}^3}{15.4 G^2 m_* \rho_{\mathrm{h}} \ln \Lambda}, \quad (23)$$

where $\rho_{\mathrm{h}} = 3 M_{\mathrm{cl}} / 8\pi R_{\mathrm{h}}^3$ is the mean density within R_{h}, and $V_{\mathrm{h}}^2 = V_{\mathrm{m}}^2(R_{\mathrm{h}})$. Heavy stars sink to the center of the cluster in a time $\ll t_{\mathrm{rh}}$.

4. Methods of Computing the Evolution of a Globular Cluster

Three different numerical methods of calculating the detailed evolution of a globular cluster have been used extensively. A cluster contains too many stars for direct Nobody techniques to be applied. Hénon (1965) developed an "orbit-averaged" Monte Carlo technique, which follows the evolution of the energy E and angular momentum J of ∼1000 test stars. Since the exact positions and velocities of the test stars are not known, statistically significant results are obtained by averaging over many orbits. Shapiro and Marchant (1976) modified Hénon's method to account for the loss of stars from the cluster. The more direct Monte Carlo method used by Spitzer and Thuan (1972) computes the positions and velocities of the test stars. Subsequently Cohn (1979, 1980) has used a variant of the orbit-averaged Fokker–Planck equation adopted from plasma physics. In this case, the averaging is done instantaneously over all particles with the same angular momentum. Goodman (1983a) has modified this approach to allow for strong encounters between stars, and has developed a Fokker–Planck code for an axisymmetric stellar system, in which the average is over particles of identical E and J_z (1983b).

5. Early Evolution of an Isolated Globular Cluster

In this section, we will review some well-established results concerning the evolution of a globular cluster before core collapse.

The prevailing idea as to how a globular cluster first attains its equilibrium state, is that the stars undergo *violent relaxation*. During the initial collapse from an unrelaxed configuration, the energy of a single star is *not* conserved, even neglecting two-body encounters, since the smoothed gravitational potential is time-dependent. Hénon (1964) and Lynden–Bell (1967) have shown that, during the collapse, the random velocities of the stars in the inner parts of the cluster may achieve a nearly Maxwellian distribution. This idea has its most important application in models of galaxy formation. The two-body relaxation time is longer than the age of the universe, in elliptical galaxies and the spheroidal component of spirals, and in clusters of galaxies; but the random velocities in these systems are observed to have an approximately Maxwellian distribution.

Since violent relaxation occurs over a dynamical timescale (that is, over only a small fraction of the two-body relaxation time) the phase space distribution function f evolves according to

$$\frac{\partial f}{\partial t} - \nabla \widetilde{\phi} \cdot \frac{\partial f}{\partial \mathbf{v}} + \mathbf{v} \cdot \frac{\partial f}{\partial \mathbf{r}} = 0, \tag{24}$$

where $\widetilde{\phi}$ is the *smoothed* potential. As the orbital phases of neighboring elements of the initial phase space distribution lose coherence, the velocity distribution approaches a Maxwellian. Since the acceleration of a star in the potential $\widetilde{\phi}$ is independent of its mass, the final random velocities of light and heavy stars are the same. The "temperature" of the Maxwellian distribution for each stellar component is proportional to the stellar mass.

One clear observational fact is that most globular clusters are nearly spherical, with ellipticities < 0.2. This fact has a straightforward explanation in the evaporation of stars from the cluster (Agekian 1958; Shapiro and Marchant 1976; Goodman 1983b). The idea that cluster rotation is damped by evaporation is supported by the fact that (presumably younger) clusters in the Magellenic clouds have larger ellipticities and rotation rates.

If a cluster rotates slowly, then an escaping star on average carries away a specific angular momentum $J \sim R_h V_{esc} \gg R_h V_{rot}$, where V_{rot} is the rotational velocity of cluster stars. The loss of stars in this part of phase space is replenished in a relaxation time. Modeling a cluster as a Maclaurin ellipsoid, Agekian (1958) and Shapiro and Marchant (1976) found that it became rounder if the initial ellipticity was less than 0.74.

Goodman (1983b) treated this problem more carefully with a numerical integration of the Fokker–Planck equation. He found that the ellipticity was damped in a few t_{rh}. This work implies that the approximation of a globular cluster as a Maclaurin ellipsoid is not quantitatively accurate, because a cluster is highly inhomogenous.

There is an instability which sets in when too large a fraction of the mass of a globular cluster is in massive stars ($cf.$ Spitzer 1987). As we discussed in Sec. Ib, two-body encounters tend to equalize the random energies of stars, transferring energy from massive stars to light stars. The random velocities of massive stars steadily decrease with time, and they tend to accumulate at the center of the cluster. The further details of this process depend on the relative fraction of the cluster mass in heavy and light stars. Consider the case of two populations with masses m_1 and $m_2, m_2 > m_1$. The total mass in each population is denoted by M_1 and M_2. If M_1 is small compared to M_2, then all the light stars are quickly expelled from the cluster. If M_2 is sufficiently small compared to M_1, then the heavy stars settle to the center of the cluster, but the central potential is still dominated by the light stars, and the two populations reach equipartition. If, however, the central potential becomes dominated by heavy stars before equipartition is reached, then the density of heavy stars in the core continues to increase, along with their kinetic energy. Eventually all but a few of the light stars are expelled from the core, and energy transfer ceases. The boundary between these two domains of behavior is where $M_1 r_{h1}^{-3} \sim M_2 r_{h2}^{-3}$ at equipartition ($GM_1 m_1/r_{h1} \sim GM_2 m_2/r_{h2}$). That is, the instability occurs for

$$\frac{M_2}{M_1} > \kappa \left(\frac{m_2}{m_1}\right)^{-3/2}. \tag{25}$$

A detailed calculation gives $\kappa = 0.16$ (Spitzer 1987).

The last, and perhaps most important, instability to consider is the *gravithermal collapse* (Antonov 1962; Lynden-Bell and Wood 1968). A self-gravitating isothermal gas sphere, contained by a rigid shell, may attain a stable hydrostatic equilibrium only when the ratio of central to outer densities is less than 710. When instability is reached, heat is conducted from the inner parts of the sphere to the outer parts; the inner parts become more compact, and the outer parts more rarefied. Thus, we expect that when the central density ρ_0 in a globular cluster exceeds the mean density ρ_h within R_h by a comparable value, the cluster core begins to contract. This contraction progresses on a timescale somewhat longer than the initial relaxation time, namely $t_{col} \simeq 14 t_{rh}$. The numerical solution of the Fokker–Planck equation yields (Cohn 1980)

$$\rho_0(t) \propto [R_c(t)]^{-2.23}, \tag{26}$$

$$\rho_0(t) = \rho_0(t=0) \left(1 - \frac{t}{t_{col}}\right)^{-1.2}, \tag{27}$$

and

$$V_{m,0}^2(t) = V_{m,0}^2(t=0) \left(1 - \frac{t}{t_{col}}\right)^{-0.12}. \tag{28}$$

At any given time during the self-similar collapse, $t - t_{col} \simeq 300 \, t_{rh,0}$. These results agree well with those obtained by Lynden-Bell and Eggleton (1980), who modeled a cluster as a continuous, conductive fluid. As the core collapses, the core mass $\sim \rho_0(t) R_c^3(t)$ gets progressively smaller, as does the outward energy flow. In other words, a relatively modest source of heat may reverse the collapse in its late stages. Hénon (1965) was the first to postulate that the formation of binaries might provide the necessary heating. He calculated self-similar solutions for a cluster expanding after core collapse, based on the Fokker–Planck equation (1961, 1965).

The process of core collapse has a real if not quite perfect analogy with stellar evolution. When the source of hydrogen fuel is exhausted in a high mass star, the helium core becomes isothermal. The core mass grows until it exceeds the Schönberg–Chandrasekhar limit, at which point the core contracts until a new energy source is tapped. Similarly, a globular cluster grows more centrally condensed until its mass exceeds the Antonov limit, at which point it undergoes core collapse. If the energy source which halts the contraction of the core in a high mass star is centrally concentrated, then the resulting Cowling models are the stellar analogs of the Hénon models.

6. Effects of Finite Stellar Size: Heating and Core Bounce

If all stars were true point masses, then two-body encounters in the core of a globular cluster would never result in the formation of binaries, since these encounters would be elastic. That is, binaries form in two-body collisions only due to the effects of *finite* stellar size. Stellar oscillations provide a sink for the kinetic energy of the passing stars.

The cross section for two stars of equal mass m_* and relative velocity V_∞ to collide is

$$\sigma = \pi R_{\min}^2 \left(1 + \frac{4Gm_*}{R_{\min} V_\infty^2} \right) \simeq 2\pi R_{\min} R_* \left(\frac{V_{R*}}{V_\infty} \right)^2 . \tag{29}$$

Here, R_* is the radius of the stars and V_{R*} is the surface escape velocity. Gravitational focussing dominates the cross section because $V_{R*} \gg V_\infty$. The minimum separation must be $R_{\min} \le R_*$ if the stars physically collide; and $R_{\min} < 2Gm_*/V_\infty^2$ is the separation required for a strong encounter, in the sense that each star is strongly deflected. Therefore the rate for strong encounters greatly exceeds that of actual collisions, by a factor $(V_{R*}/V_\infty)^2$.

Binary formation is a rare process, in the sense that a two-body encounter results in a binary only when the minimum separation is comparable to the size of the stars. Consider two stars approaching each other from a large distance, with relative velocity $V_\infty = 2V_m$. Each star excites tides adiabatically on the other, and the energy dissipated in each star during the encounter is (Fabian, Pringle and Rees 1975; Press and Teukolsky 1977; Lee and Ostriker 1986)

$$\Delta E = 5 \frac{Gm_*^2}{R_*} \left(\frac{R_*}{R_{\min}} \right)^{10} . \tag{30}$$

The stars capture each other if $\Delta E > \frac{1}{2} m_* V_\infty^2$, or

$$R_{\min} \le 1.26 \left(\frac{V_{R*}}{V_\infty} \right)^{0.2} \simeq 2.5 R_* . \tag{31}$$

The net capture cross-section is

$$\sigma_{2b} = f \cdot \pi R_*^2 \left(\frac{V_{R*}}{V_\infty} \right)^2 ; \quad f = 0.54 \left(\frac{V_{R*}}{V_\infty} \right)^{0.2} . \tag{32}$$

The rate of two-body binary formation in a cluster core is thus

$$\nu_{2b} \cdot t_{\rm rh} = 0.24 \frac{\ln \Lambda}{f} \left(\frac{V_{R*}}{V_\infty} \right)^2 \simeq 1200 \left(\frac{V_{R*}}{600 \, \text{km s}^{-1}} \right)^{2.2} \left(\frac{V_m}{20 \, \text{km s}^{-1}} \right)^{-2.2} . \tag{33}$$

During core collapse, however, the rate of binary formation increases as $\nu_{2b}(t) \propto \rho_0(t)^{-1.1}$; and the effective relation time $t_{\rm rh,eff} \sim \rho_0(t)/\rho_0(t) \sim 3.10^{-3} t_{\rm rh,0}$, where $t_{\rm rh,0}$ is the central relaxation time. As as result, binaries are formed faster than they are ejected from the core when ρ_0 has increased by a factor of $\sim 10^6$. An integration of the Fokker–Planck equation gives for the number of core binaries (Statler, Ostriker and Cohn 1987)

$$\frac{N_{\rm b}}{N_*} = \frac{30}{\ln \Lambda} \left(\frac{V_m}{V_{R*}} \right)^{1.8} \left[1 - \left(\frac{\Delta t}{t_{\rm col}} \right)^{0.29} \right] , \tag{34}$$

where the time is measured from the onset of core collapse, $\Delta t = t - 9t_{\rm rh,0}$. In a cluster with $N_* = 10^6$ and r.m.s. velocity $V_m = 20 \, \text{km s}^{-1}$, the number of binaries formed from $t = 9t_{\rm rh,0}$ to collapse is $N_{\rm b} = 3000$. From $t = 6t_{\rm rh,0}$ to $9t_{\rm rh,0}$ another

2000 binaries are formed, and the fraction of stars in binaries (formed by two-body interactions) is

$$\frac{N_b}{N_*} \sim 5 \times 10^{-3}. \tag{35}$$

This number of binaries will heat the core sufficiently to reverse the collapse.

The binaries formed by this mechanism are so tight, that the stellar orbits are circularized in $<10^7$ yr. The resulting orbital radius $R_{circ} = 2R_{min} \simeq 5R_*$. We conclude that two-body encounters produce only very close binary pairs, which decay by radiating gravitational waves in $\sim 10^9$ yr.

Binaries are also formed in three-body interactions. Let us determine their rate of formation in rough order of magnitude. The density of pairs of stars with mutual separation less than Δr is $\sim n_*^2 (\Delta r)^3$, and the collision rate of stars with each of these pairs is $\sim n_*^3 (\Delta r)^2 V_m$. Now a binary will not form unless $\Delta r \sim Gm_*/V_m^2$, or smaller, so that the overall rate is $\sim n_*^3 G^3 m_*^3 / V_m^9$. A detailed calculation of the overall rate (Hut 1985) gives

$$\nu_{3b} = 0.9 \frac{n_*^3 G^3 m_*^3}{V_m^9}. \tag{36}$$

A binary is expected to form only rarely (in a relaxation time) since

$$\nu_{3b} \cdot t_{rh} = \frac{0.4}{f N_*^2} \left(\frac{V_{R*}}{V_m} \right)^2, \tag{37}$$

where f is defined in Eq. (32). It seems that only a few three-body binaries form during core collapse.

Let us estimate the rate at which stars in the core are heated by interactions with binaries. If V_∞ is the relative velocity between a binary and a third star, each with masses m_b and m_* ($m_* < m_b$), then for an impact parameter less than the binary separation a, the energy imparted to the third star is $\Delta E \sim m_*(Gm_b/a)$. The cross section for this interaction is dominated by gravitational focussing, $\sigma \simeq 2\pi a G(m_b + m_*)/V_\infty^2$, so the heating rate per unit volume is

$$\dot{E}_b = \Delta E \cdot n_b n_* \sigma V_\infty \sim \rho_* \frac{G^2(m_b + m_*)}{V_\infty} n_b. \tag{38}$$

One may readily see that this heating rate will halt core collapse, even for a small number of binaries. However, the calculations of Spitzèr and Mathieu (1980) show that the binaries are simultaneously ejected when they scatter off stars in the core. Thus, binaries will prevent core collapse only if their rate of formation is sufficiently rapid.

In a series of numerical experiments, Statler, Ostriker and Cohn (1987) have investigated the collapse, bounce and re-expansion of a globular cluster core, as a result of heating by the formation of tidally captured binaries. The experiments calculate the interactions between stars in the Fokker–Planck approximation, and use a Plummer model for the initial equilibrium state. During core collapse, the energy dissipated in the process of forming binaries is accounted for, as are the effects of dynamical friction. Heating and ejection of core stars occurs as a result of interactions both between single stars and binaries, and between two binaries, leading to the halt of core collapse and re-expansion of the cluster. The effect of the ejection of core stars on the stars remaining in the core is self-consistently accounted for.

The following simplifications were made in these experiments. The stars were assumed to have constant mass, and a locally isotropic velocity distribution. The merging of field stars in physical collisions, and the decay of binary orbits due to gravitational radiation, were neglected. Finally, no account was made for stellar evolution.

The initial state chosen for the cluster is a reasonable representation during one or two relaxation times after formation. As lighter stars evaporate from the cluster, and heavier stars settle to the core, the cluster passes through a sequence of isothermal equilibria. Eventually, it reaches an unstable equilibrium, and by $t \simeq 14t_{\rm rh}$ the core has collapsed. The sudden growth of the central density during core collapse, and the subsequent bounce, are shown in Fig. 1 (from Statler *et al* 1987). During re-expansion, the central density $\rho_0 \propto t^{-4/3}$, the central velocity dispersion $(V_{\rm m}^2)_0 \propto t^{-2/3}$ and the half-mass radius $R_{\rm h} \propto t^{2/3}$. These scaling relations were first obtained by Hénon (1961, 1965). In addition, the core radius $R_{\rm c} \propto t^{1/3}$. Three-body binary formation is always negligible compared to tidal capture. The density

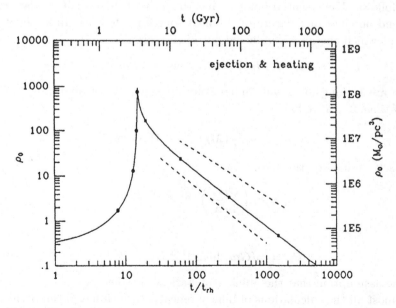

Fig. 1.   Total central density as a function of time. From Statler *et al* (1987).

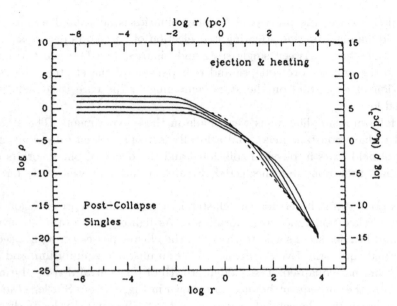

Fig. 2.   Density profile of single stars at the times labeled by squares in Fig 1. From Statler *et al* (1987).

profile through the core as a function of time is shown in Fig. 2 during collapse and re-expansion.

We may obtain the above scaling relations for the quasi-static expansion of the core, as follows. The essential point is that the form of these scaling relations does not depend on the exact nature of the heat source in the core. In a steady state, the rate at which energy is conducted out of the core is

$$L_{\text{core}} = \text{const} \times M_h V_h^2 / t_{\text{rh}} = \text{const} \times V_h^5 m_* \ln \Lambda / G M_h.$$

Now, we are looking for a self-similar solution in which all quantities scale as a power of time. Then we have

$$L_{\text{core}} \sim \frac{d}{dt} \left(M_h V_h^2 \right) \sim M_h V_h^2 / t$$

and the self-similar solutions are

$$V_h = k_1 \left(\frac{G m_*}{\ln \Lambda} \right)^{1/3} N_*^{2/3} \, t^{-1/3}, \tag{39}$$

and

$$R_h = k_2 (G m_* \ln \Lambda^2)^{1/3} \, N_*^{-1/3} \, t^{2/3} \tag{40}$$

under the assumption that the number of stars N_* is constant.

In almost all the calculations of binary reheating which have been performed to date, the effects of physical collisions and mergers between stars have been neglected.

While the calculations of Lee and Ostriker (1986) indicate that a significant fraction of tidally captured binaries may in fact be mergers (a very close encounter is required to bind the stars), the effects of mergers on the overall evolution of a cluster are probably small. Lee (1987) has studied core bounce under the assumption that *all* two-body binaries end up as mergers. He finds that for $10^5 < N_* < 10^6$, the massive stars formed in subsequent mergers evolve quickly and eject their mass from the core. The heating of the core is dominated by this process. The important point here is that the rate of mass loss from the core is fixed by the rate at which binaries form by tidal capture; how the mass is lost is only of secondary importance to the global evolution of the cluster. Three-body binaries are the dominant heat source only for $N_* < 10^5$.

7. External Effects on the Evolution of a Globular Cluster

A globular cluster passing near or through another object, such as the Galactic disk or bulge, or the Magellenic clouds, is subject to a "tidal shock". In the center-of-mass frame of the cluster, each star experiences a time-dependent tidal field. In the outer parts of the cluster, some stars may receive enough energy to escape.

The simplest problem to consider is a cluster passing a point mass $\mathcal{M}$, with velocity V_{cl} and distance of closest approach Δr. The tidal acceleration at R_{h} is $\sim G\mathcal{M}R_{\text{h}}(\Delta r)^{-3}$, so that a typical star receives a kick $\Delta V \sim G\mathcal{M}R_{\text{h}}(\Delta r)^{-2}V_{\text{cl}}^{-1}$ during the time of closest approach $\sim \Delta r/V_{\text{cl}}$. The cluster will be disrupted if $(\Delta V)^2 > V_{\text{h}}^2 \simeq GM_{\text{cl}}/2R_{\text{h}}$, where M_{cl} is the cluster mass.

The first calculation of Galactic tidal heating of a cluster was done for the case of a cluster passing through the disk (Ostriker, Spitzer and Chevalier 1975). A cluster with orbital period t_{cl} is disrupted after a time

$$t_{\text{shock}} = t_{\text{cl}} \frac{GM_{\text{cl}}V_{\text{cl}}^2}{20g^2R_{\text{h}}^3}, \tag{41}$$

where V_{cl} is the velocity of the cluster through the disk, and g is the gravitational acceleration toward the disk, exterior to the disk. The compatibility of Eq. (41) with the estimate above may be seen by the substitution $g \rightarrow G\mathcal{M}/(\Delta r)^2$.

Another important effect to consider is the loss of stars from a cluster which is at rest in a static but inhomogenous external gravitational field. Recall that stars in the exponential tail of the velocity distribution leave the cluster, and other stars fill their orbits only over a relaxation time. The rate of mass loss from a post-collapse cluster is therefore

$$\frac{dM_{\text{cl}}}{dt} = -\alpha \frac{M_{\text{cl}}}{t_{\text{rh}}}, \tag{42}$$

where the constant $\alpha \simeq 4 \times 10^{-3}$ is small because the number of stars with $V^2 > \langle V_{\text{esc}}^2 \rangle = 4\langle V_m^2 \rangle$ is small. Indeed, the rate of mass loss is very nearly constant, as the following argument shows. Substituting expression (23) for t_{rh} in Eq. (42), and using the fact that the ratio $t_d(R_t)/t_d(R_h)$ is constant during self-similar expansion, we find

$$t_{evap} = \frac{1}{\alpha} t_{rh} \simeq \frac{N_*}{\ln N_*} t_d(R_h) \propto \frac{N_*}{\ln N_*} t_d(R_t) \simeq \frac{N_*}{\ln N_*} \cdot \frac{R_{cl}}{V_{cl}},$$

or

$$t_{evap} = 1.3 \times 10^{10} \left(\frac{N_*}{10^5}\right) \left(\frac{R_{cl}}{1\,\text{kpc}}\right) \left(\frac{V_{cl}}{220\,\text{km s}^{-1}}\right)^{-1} \text{ yr.} \qquad (43)$$

Here we have also used the fact that the dynamical time of a cluster star at the tidal radius is comparable to the orbital time of the cluster through the Galaxy, R_{cl}/V_{cl}. We conclude that the cluster mass decreases linearly with time, $M_{cl} = M_{cl}(0)(1 - t/t_{evap})$.

A repeat of the simulations of core bounce of Statler et al. (1987), with the effects of the Galactic tidal cut-off included, gives very similar results (Ostriker and Lee 1987). During the re-expansion of the cluster after core collapse, stars flow over the tidal boundary are lost from the cluster.

A globular cluster moving through the Galaxy suffers dynamical friction in a similar manner to a single star moving through a field of stars. The cluster's orbit decays, and it spirals toward the Galactic center. As calculated by Tremaine et al. (1975), the timescale for this to occur is

$$t_{df} = 8 \cdot 10^9 \left(\frac{N_*}{10^5}\right)^{-1} \left(\frac{R_{cl}}{10\,\text{kpc}}\right) \left(\frac{V_{cl}}{220\,\text{km s}^{-1}}\right) \text{ yr.} \qquad (44)$$

That is, massive clusters within a distance $R_{cl} = 1\,\text{kpc}$ of the Galactic center suffer significant decay of their orbits.

The potential of the Galaxy is almost certainly *triaxial*, rather than perfectly axisymmetric. As a result, the motion of a cluster is never exactly radial, and its orbit precesses. Some fraction of halo orbits have envelopes with the shape of a pinched box, hence the term *box orbit*. The number of times a cluster has passed near the Galactic center in the age t_g of the Galaxy is

$$N \sim t_g \sigma/2R_a = 750(t_g/10^{10}\,\text{yr})(\sigma/150\,\text{km s}^{-1})(R_a/1\,\text{kpc})^{-1}, \qquad (44a)$$

where σ is the radial velocity of the cluster, and R_a is the apogalacticon of the cluster orbit. If the minimum separation of the long side of the box from the Galactic center is ωR_a, then the distance of closest approach of the cluster to the Galactic center is

$$R_{min} \sim N^{-1/2}\omega R_a$$

$$= 15 \left(\frac{\omega}{0.4}\right) \left(\frac{R_a}{1\text{kpc}}\right)^{3/2} \left(\frac{t_g}{10^{10}\,\text{yr}}\right)^{-1/2} \left(\frac{\sigma}{150\,\text{km s}^{-1}}\right)^{-1/2} \text{ pc.} \qquad (45)$$

As a result, we may place a constraint on the mass and core radius of a cluster, if it is not to be totally disrupted by the tidal shock suffered during its closest approach to the Galactic nucleus:

$$\left(\frac{N_*}{10^5}\right)\left(\frac{R_{\rm h}}{5\ {\rm pc}}\right)^{-3} > 0.9\left(\frac{M_{\rm n}}{10^7 M_\odot}\right)^2\left(\frac{R_{\rm a}}{1\ {\rm kpc}}\right)^{-6}. \tag{46}$$

Here, M_n is the mass of the nucleus.

The main conclusions of this section are, then, as follows. As a globular cluster collapses and its core grows denser, a heat source turns on in the core, halting the collapse of the cluster and causing it to re-expand. The main heating mechanism is the interaction of stars in the core with tight binaries formed during close two-body encounters. Since the time to core collapse is often shorter than the age of the Galaxy, one suspects that many clusters have already passed through core collapse. Ultimately, clusters evaporate totally in the tidal field of the Galaxy, this taking roughly $N_*/100$ orbits, where N_* is the original number of stars in the cluster.

8. Evolution of the Globular Cluster System: Destruction of Clusters

Let us summarize the processes which may result in the destruction of globular clusters (*cf.* Ostriker 1987).

First, many globular clusters will have already undergone core collapse, but the formation of binaries apparently heats the core and causes the cluster to re-expand. During re-expansion, stars flow over the tidal radius of the cluster, and the estimate of the previous section shows that many clusters will have lost a significant fraction of their mass in this fashion.

Globular clusters may also be disrupted by shock heating, as they pass near or through a massive object. Calculations of this effect were first done for the Galactic disk (Ostriker et al. 1975), and have been performed more recently for the Galactic bulge (Aguilar et al. 1987), and massive black holes in the halo (Wielen 1987). The heating of a cluster by giant molecular clouds (which are denser than the cluster) may be neglected compared to heating by the rest of the disk (Grindlay 1985; Chernoff et al. 1986). (See section Ij for a discussion of the observable consequences of massive black holes.) Wielen finds that a substantial fraction of clusters should have been disrupted if the halo is composed of $\sim 10^6 M_\odot$ black holes.

Chernoff et al. and Aguilar et al. have investigated the evolution of the system of globular clusters. The latter work includes most of the above effects, with the exception of shocks by GMC's. In addition the triaxiality of the Galactic potential is ignored, as are the effects of normal stellar evolution. The main conclusion reached by Aguilar et al. is that tidal shocks in the Galactic bulge are the most destructive process. Most clusters with apogalacticon $R_{\rm a} < 2\,{\rm kpc}$ are destroyed. Only dense clusters on circular orbits survive today. The inner scale radius of the globular cluster system is established in this manner. The main effect outside $R_{\rm a} = 2\,{\rm kpc}$ is that the cluster orbits are isotropized. Thus, if the initial cluster distribution was as anisotropic as the distribution of the extreme RR Lyrae stars (2 : 1 axis ratio), then loss cone effects would by now have produced an essentially isotropic distribution. A significant fraction of clusters in the range $3\,\mathrm{kpc} < R_\mathrm{a} < 10\text{--}15\,\mathrm{kpc}$ which have either:

 (i)   very eccentric orbits;
 (ii) very high mass; or
(iii) very low density, are destroyed.

In constrast, essentially all clusters survive at $R_\mathrm{a} > 15\,\mathrm{kpc}$. It is possible that a significant part of the spheroidal (or bulge) population in the inner parts of the Galaxy is the remnant of disrupted globular clusters. However, the current rate of destruction of clusters is only a few per $10^9\,\mathrm{yr}$.

   We may summarize how the evolution of a cluster is expected to depend on its mass M_cl and half-mass radius R_h. At any given R_h, there is a minimum M_cl below which the cluster will already have evaporated by tidal overflow; and a maximum M_cl above which the cluster's orbit will already have decayed due to dynamical friction. For any given M_cl, there is a minimum R_h below which the cluster will have undergone core collapse, and a maximum R_h above which tidal shocks will have disrupted the cluster. That is, there is a finite region of the $M_\mathrm{cl} - R_\mathrm{h}$ plane at a fixed distance R_cl from the Galactic center, in which clusters are expected to have survived. The area of this region decreases as R_cl decreases, and vanishes at $R_\mathrm{cl} \sim 2\,\mathrm{kpc}$.

9. Outstanding Issues

Let us now consider some unresolved questions about globular clusters. First, Peebles (1984) has suggested that globular clusters might have dark matter haloes, but there is no positive evidence in favor of this idea, and some evidence against it. Observations of the surface brightness and line-of-sight velocity dispersion in the outer parts of clusters are compatible with the presence of non-luminous matter. But the introduction of dark matter increases the mean density within R_t, so that the cut-off in star-light at R_t can no longer be explained by the Galactic tidal field.

   In the presence of dark matter, core collapse happens sooner, and clusters sink faster to the center of the Galaxy due to the effects of dynamical friction. Our understanding of these processes is not sufficiently quantitative to place any interesting constraints on the presence of dark haloes.

   A second issue is: what fraction of the initial population of globular clusters has been destroyed? Within $2\,\mathrm{kpc}$ from the center of the Galaxy, this is probably a large fraction; it is possible that the bulge is composed of the remains of old clusters. Outside $2\,\mathrm{kpc}$, this fraction is very uncertain. Most destruction mechanisms do not work efficiently in this region — but only for clusters of observed densities. It is possible that an initial population of low density clusters has been disrupted.

   Third, it is not clear whether there is any evidence for triaxiality of the Galactic potential in the distribution of clusters.

Fourth, globular clusters are a probe of the distribution of mass in the outer part of the Galaxy. Tremaine and Lee (this volume) discuss how the outer clusters as well as the dwarf satellites may be used to place constraints on the mass distribution in the outer parts of the Galaxy. Since the line-of-sight velocity of a cluster is the only dynamical information available, the statistical uncertainties are considerable.

Finally, globular clusters have a potential use as a cosmological distance indicator. Unlike galaxies, globular clusters have a well-defined peak to their luminosity function, and the characteristic luminosity at the peak is similar in the Galaxy and in M31. The width of the peak is, however, different in M87. The position of the peak is not measured accurately enough in other galaxies to make this method immediately useful.

10. Nature of the Dark Halo of Our Galaxy

It is possible that most of the matter in the halo of our galaxy is baryonic. Recall that the "local" mass-to-light ratio $dM/dL = dM/dR \cdot (dL/dR)^{-1}$ grows from $\simeq 5$ at the center of the Galaxy, to $\simeq 500$ at $R = 15$ kpc. Clearly, in the outer parts of the halo, there is a significant amount of unseen material. The calculations of primordial nucleosynthesis imply that the mean cosmic density of baryons, as a fraction of closure density is (see, for example, Yang $et\ al$ 1985), $\Omega_b = (0.04\text{--}0.14)h_{50}^{-2}$ where h_{50} is Hubble's constant in units of $50\,\mathrm{km\,s^{-1}\,Mpc^{-1}}$. In comparison, the fraction of closure density observed in stars is at most $\Omega_* = 0.01$, and indirect arguments imply that all components of the inter-galactic medium contribute $\Omega_{IGM} = 0.03$ (Ikeuchi and Ostriker 1986). Therefore, if the calculations of primordial nucleosynthesis correctly predict the cosmic baryon abundance, and if h_{50} is closer to 1 than 2, then most baryons must exist in the form of dark matter. It is an interesting coincidence that all dynamical measurements of the density of the universe, which are sensitive to dark matter in galactic haloes and in clusters of galaxies, give values for Ω in the range 0.1–0.3 (see, for example, Peebles 1986).

If most of the the Galaxy's mass, including the dark matter, is baryonic, we may ask what form it may take. Most of the mass in the Galactic halo *cannot* be gaseous. If it were hot and hydrostatically supported ($\sim 10^7$ K), the emission at ~ 1 ke V would significantly exceed the brightness of the soft X-ray background. If it were in cool lumps ($\sim 10^4$ K), it would again be too bright, in this case at 21 cm. It is also difficult to see how most of the mass could be in dust, since one would expect at least 100 times as much H and He to be present.

Low mass stars are certainly a viable candidate for the main constituent of the halo, as long as they are light enough ($<0.07 M_\odot$) that nuclear burning does not occur in their cores. One might worry that if most of the stars had masses just below this limit, then there would be a hint of their presence in the mass function of Population II objects.

There is no firm objection to most baryons being incorporated in stars that form before galaxies, or "Population III" objects. These could be clusters of low mass stars with masses $\sim 10^6 M_\odot$ (the Jeans mass at decoupling; *cf.* Peebles and Dicke 1968, and Fall and Lacy 1986), or supermassive stars that collapsed to form black holes. However, black holes with masses in the range $10\text{--}100 M_\odot$ can probably be excluded, because the stars that collapsed to form them would over-produce heavy elements (Carr and Rees 1984). For example, a $100 M_\odot$ star forms a black hole of $\sim 20 M_\odot$ and ejects $\sim 20 M_\odot$ of material with $Z > 6$, as well as $\sim 60 M_\odot$ of helium and unprocessed hydrogen. Supermassive stars ($\gg 10^2 M_\odot$) collapse directly to a black hole without ejecting any material; so we cannot rule out supermassive black holes in a similar manner.

Let us consider some of the consequences of the existence of supermassive black holes in the halo of the Galaxy (Lacey and Ostriker 1985). The most interesting observable effect is that the black holes will heat stars as they pass through the disk, causing it to flair. One may immediately predict that older stars have a larger scale height, since they have been heated longer.

The expected increase in the vertical velocity of a disk star is, per unit time,

$$\left\langle \frac{\partial}{\partial t} \Delta V_\perp^2 \right\rangle = 2^{1/2} 4\pi \frac{G^2 n_H m_H^2}{\sigma_H} \ln \left[\Lambda \cdot \frac{\Phi(x) - G(x)}{x} \right] \equiv 4 D_z, \qquad (47)$$

where σ_H is the one-dimensional velocity dispersion of the black holes, m_H their mass, and n_H their density in the halo. The dimensionless variable $x = \sqrt{V_*^2 / 2\sigma_H^2}$, where V_* is the velocity of the disk star with respect to the black hole population. The functions Φ and G are defined in Sect. Ib.

The vertical velocity dispersion of the disk grows as

$$\sigma_z^2(t) = \left(\sigma_z^2(0) + D_z t \right)^{1/2}, \qquad (48)$$

since in interactions between a light stellar component and a heavier, more energetic component, the light component is heated. This result matches with the vertical profile of the disk of the Galaxy, as well as other spirals, if $m_H \simeq 2 \times 10^6 M_\odot$. In particular, the time-dependence $\propto t^{1/2}$ fits the observed distribution of scale heights in stars of different ages (Wielen 1987). The value for $V_\perp^2 / V_\parallel^2$ obtained also agrees well with that observed in the Galaxy.

To test this model, we may ask what it predicts for the high velocity tail of the disk population — in particular, of a group with low velocity dispersion. The distribution of vertical velocities w (normalized to one) has the form

$$\begin{aligned} f(w) &= \frac{1}{\sqrt{2\pi}\sigma_w} \exp\left(-\frac{w^2}{2\sigma_w^2} \right), \quad w < w_c, \\ &= \frac{\sqrt{\pi}}{\ln \Lambda} \frac{\sigma_D \sigma_w^3}{w^5}, \quad w_c < w < w_H, \end{aligned} \qquad (49)$$

where σ_D is the vertical velocity dispersion of the background disk, and $w_c \simeq 3.6 \sigma_w$. Now, there is a population of high velocity A stars with normal disk metallicities, extending several kiloparsecs from the disk, with $\sigma_w = 10\,\mathrm{km\,s^{-1}}$. A fraction $\sim 10^{-3}$ of A stars are observed to have $w > 50\,\mathrm{km\,s^{-1}}$; the prediction of Eq. (48) is 3×10^{-4}.

References

Agekian, T. A. (1958) *Sov. Astron. A.J.* **2**, 22.

Aguilar, L., Hut, P., and Ostriker, J. P. (1987) In preparation.

Ambartsumian, V. A. (1938) *Ann. Leningrad State Univ., No. 22.*

Antonov, V. A. (1962) Translated in: *Dynamics of Star Clusters, IAU Symposium No. 113*, ed. Goodman, J. and Hut, P., Reidel, Dordrecht, The Netherlands.

Binney, J. and Ostriker, J. P. (1987) In preparation.

Carr, B. J. and Rees, M. J. (1984) *Mon. Not. R. Astron. Soc.* **206**, 315.

Chernoff, D. F., Kochanek, C. S., and Shapiro, S. S. (1986) *Astrophys. J.* **309**, 183.

Cohn, H. (1979) *Astrophys. J.* **234**, 1036.

Cohn, H. (1980) *Astrophys. J.* **242**, 765.

Fabian, A. C., Pringle, J. E., and Rees, M. J. (1975) *Mon. Not. R. Astron. Soc.,* **172**, 15p.

Fall, S. M. and Lacey, C. G. (1986) Preprint.

Goodman, J. (1983a) *Astrophys. J.* **270**, 700.

Goodman, J. (1983b) *Princeton Thesis.*

Goodman, J. (1984) *Astrophys. J.* **280**, 298.

Grindlay, J. E. (1985) In: *Dynamics of Star Clusters, IAU Symposium No. 113*, p. 43, ed. Goodman, J. and Hut, P., Reidel, Dordrecht, The Netherlands.

Hénon, M. (1961) *Ann. d'Astroph.* **24**, 179.

Hénon, M. (1964) *Ann. d'Astroph.* **27**, 83.

Hénon, M. (1965) *Ann. d'Astroph.* **28**, 62.

Hut, P. (1985) In: *Dynamics of Star Clusters, IAU Symposium No. 113*, p. 231, ed. Goodman, J. and Hut, P., Reidel, Dordrecht, The Netherlands.

Ikeuchi, S. and Ostriker, J. P. (1986) *Astrophys. J.* **301**, 522.

King, I. (1966) *Astron. J.* **71**, 64.

Lacey, C. G. and Ostriker, J. P. (1985) *Astrophys. J.* **299**, 633.

Lee, H. M. (1987) *Astrophys. J.* **319**, 801.

Lee, H. M. and Ostriker, J. P. (1986) *Astrophys. J.* **310**, 176.

Lynden-Bell, D. (1967) *Mon. Not. R. Astron. Soc.* **136**, 101.

Lynden-Bell, D. and Eggelton, P. (1980) *Mon. Not. R. Astron. Soc.* **191**, 483.

Lynden-Bell, D. and Wood, R. (1968) *Mon. Not. R. Astron. Soc.* **138**, 495.

Oort, J. H. (1977) *Astrophys. J. Lett.* **218**, L97.

Ostriker, J. P. (1987) In: *The Harlow Shapley Symposium on Globular Cluster Systems in Galaxies, IAU Symposium No. 126*, ed. Grindlay, J. E. and Phillips, A. G. D., Reidel, Dordrecht, The Netherlands.

Ostriker, J. P. and Lee, H. M. (1987) *Astrophys. J.*, in press.

Ostriker, J. P., Spitzer, L., and Chevalier, R. A. (1973) *Astrophys. J. Lett.* **176**, L51.

Peebles, P. J. E. (1984) *Astrophys. J.* **277**, 470.

Peebles, P. J. E. (1986) *Nature* **321**, 27.

Peebles, P. J. E. and Dicke, R. H. (1968) *Astrophys. J.* **154**, 891.

Press, W. H. and Teukolsky, S. A. (1977) *Astrophys. J.* **213**, 183.

Shapiro, S. L. and Marchant (1976) *Astrophys. J.* **210**, 757.

Spitzer, L. (1987) Princeton University Press, Princeton.

Spitzer, L. and Hart, M. H. (1971) *Astrophys. J.* **164**, 399.

Spitzer, L. and Mathieu, R. D. (1980) *Astrophys. J.* **241**, 618.

Spitzer, L. and Thuan, T. X. (1972) *Astrophys. J.* **175**, 31.

Tremaine, S., Ostriker, J. P., and Spitzer, L. (1975) *Astrophys. J.* **196**, 407.

Wielen, R. (1987) In: *The Harlow Shapley Symposium on Globular Cluster Systems in Galaxies, IAU Symposium No. 126*, ed. Grindlay, J. E. and Phillips, A. G. D., Reidel, Dordrecht, The Netherlands.

Yang, J., Turner, M. S., Steigman, G., Schramm, D. N., and Olive, K. A. (1984) *Astrophys. J.* **281**, 493.

Zinn, R. (1985) *Astrophys. J.* **293**, 424.

Chapter 5

POSITIVE ENERGY PERTURBATIONS IN COSMOLOGY – II

J. P. Ostriker

Princeton University Observatory
Princeton, NJ 08544, USA

C. Thompson*

Joseph Henry Laboratories
Princeton, NJ 08544, USA

The purpose of this second part is to study the effects of a localized injection of energy in an expanding universe. We are mainly interested in cases where hydrodynamical forces are important. A far more detailed treatment is presented in Ostriker and McKee (1988).

1. Hydrodynamics in a Cosmological Setting

To begin, let us review the properties of a compressible, inviscid fluid (see, for example, Landau and Lifshitz 1959). We shall be satisfied with a completely non-relativistic treatment, and treat only perturbations small compared to the horizon. The equation of continuity is

$$\frac{\partial \rho}{\partial t} + \nabla \cdot (\rho \mathbf{u}) = 0, \tag{1}$$

Euler's equation is

$$\frac{\partial \mathbf{u}}{\partial t} + (\mathbf{u} \cdot \nabla)\mathbf{u} = -\frac{1}{\rho}\nabla P - \nabla \phi + \mathbf{f}, \tag{2}$$

and the equation expressing conservation of energy is

$$\frac{\partial}{\partial t}\left(e + \frac{1}{2}u^2 + \phi\right) + \nabla \cdot \left[\rho \mathbf{u}\left(h + \frac{1}{2}u^2 + \phi\right)\right] = 0. \tag{3}$$

Here, ρ is the density of the fluid and P its pressure. The specific enthalpy $h = \frac{\gamma}{\gamma-1}(kT/\mu)$ for a perfect gas with ratio of specific heats γ and mean molecular weight μ. The gravitational potential ϕ is calculated in the Newtonian approximation; for a perturbation with spherical symmetry in a homogeneous, isotropic

---

*Current address: CITA, University of Toronto, Toronto, ON M5S 3H8, Canada.

universe,

$$-\nabla\phi = -\frac{GM(<r)}{r^2} \tag{4}$$

where $M(<r) = 4\pi \int_0^r [\rho(r) + 3P(r)/c^2] r^2 dr$ is the gravitational mass interior to r. The additional acceleration $\mathbf{f}$ include effects such as inverse Compton drag between electrons and the cosmic background radiation. The velocity $\mathbf{u}$ is measured with respect to the center of symmetry, and may be written

$$\mathbf{u} = \mathbf{v} + H\mathbf{r}, \tag{5}$$

when $r \ll ct, u \ll c$. The second term is the velocity of the unperturbed Hubble flow, and $\mathbf{v}$ is the "peculiar" velocity with respect to this flow. If the universe is flat, and dominated by non-relativistic matter, then the unperturbed density $\rho(t) \propto r^{-3}(t)$, and the solution to Eq. (1) and (2) is $r \propto t^{2/3}$ and $\rho = (6\pi Gt^2)^{-1}$. It follows that Hubble's constant $H = \dot{r}/r = 2/(3t)$. We will usually take the fractional density in baryonic matter to be $\Omega_b < 1$, and assume that the remaining component — the *dark matter* — interacts only gravitationally. (It should be remembered that much of the dark matter for which there is dynamical evidence may be baryonic, and that only one measurement, that of Loh and Spillar (1986), requires the existence of non-baryonic dark matter.)

We adopt a point of view suitable for the study of the dirty details of (spherical) perturbations to the Hubble flow, in which the "center" of the expansion lies at the center of the perturbation. The Hubble flow at any radius is braked by the gravitational attraction of matter with that radius. Now, if we completely remove all matter within a sphere, then Birkhoff's theorem tells us that the spacetime inside this sphere is flat. In other words, we may calculate the gravitational acceleration a shell (of mass ΔM and radius R) by assuming that it moves in an empty Euclidean space and feels only the gravitational attraction of the material within it. The potential energy with respect to infinity of the shell (moving in empty space) is

$$-GM(<R)\Delta M$$

$$R = -\frac{1}{2}\Delta M(HR)^2,$$

and the kinetic energy of the shell is $-\frac{1}{2}\Delta M(HR)^2$. In this Newtonian approximation, the Hubble flow has *zero* net energy. This result does not surprise us, since in the limit $t \to \infty$, the flow velocity $u \to 0$. We do note, however, that even if kinetic and potential energy are presently observed to balance only roughly, they do balance to a very high degree of accuracy at early times. (We can contemplate extrapolating our classical cosmological model back to times $t \sim \sqrt{\hbar G/c^5} = 10^{-43}$ s!) The extraordinary fine tuning at early times required to produce the rough equality between kinetic and potential energy observed today is sometimes known as as the flatness problem, and was one of the main motivations for the idea of an inflationary universe (Guth 1981).

We therefore call a local release of energy — from a galaxy undergoing a burst of star formation, a quasar, or even a superconducting cosmic string — a *positive energy perturbation* to the Hubble flow. A similar class of perturbations which we will not discuss are those which start as a localized negative density fluctuation with positive potential energy relative to the unperturbed Hubble flow.

If the energy is released in a short time and a small volume, then an expanding shock wave forms. This shock quickly becomes spherical if the surrounding medium is homogeneous, since the shock moves subsonically with respect to its hot interior. Gravity further increases the tendency of blast waves to become spherical (Bertschinger 1983).

To proceed further, we require some information about shock waves. Consider the problem of time-independent flow in one dimension. Equations (1–3) reduce to

$$\frac{\partial}{\partial x}\left(\rho u\right) = 0,$$

$$\frac{\partial}{\partial x}\left(P + \rho u^2\right) = \rho\left(-\frac{\partial \phi}{\partial x} + f\right), \qquad (6)$$

$$\frac{\partial}{\partial x}\left(h + \frac{1}{2}u^2 + \phi\right) = 0.$$

An important property of these equations is that, having fixed u, ρ, P and h at one point in the flow, these variables may take on one of *two* sets of values throughout the rest of the flow. In particular, a discontintuity may exist in the flow, across which

$$\rho_1 u_1 = \rho_2 u_2,$$

$$P_1 + \rho_1 u_1^2 = P_2 + \rho_2 u_2^2, \qquad (7)$$

$$h_1 + \frac{1}{2}u_1^2 = h_2 + \frac{1}{2}u_2^2.$$

In the case of a detonation wave, this last condition is generalized to

$$h_1 + \frac{1}{2}u_1^2 + \varepsilon c^2 = h_2 + \frac{1}{2}u_2^2. \qquad (8)$$

This discontinuity is a *shock wave*. The subscript 1 will denote fluid moving toward the shock, and 2 fluid moving away. (We choose a frame in which the shock is at rest.) In a perfect gas, the jump conditions are (for $\varepsilon = 0$)

$$\frac{\rho_2}{\rho_1} = \frac{u_1}{u_2} \simeq \frac{\gamma + 1}{\gamma - 1},$$

$$P_2 \simeq \frac{2\rho_1 u_1^2}{\gamma - 1}, \qquad (9)$$

$$\frac{kT_2}{\mu} \simeq 2\frac{\gamma - 1}{(\gamma + 1)^2}u_1^2.$$

when the Mach number $u_1/c_1 \gg 1$. Here, $c = \sqrt{\gamma kT/\mu}$ is the speed of sound in a perfect gas with ratio of specific heats γ, and mean molecular weight μ.

One obtains the appropriate boundary conditions for the blast wave by substituting the shock velocity U_S for u_1, if the ambient medium is static; whereas one substitutes the peculiar velocity $V_S = U_S - HR_S$ if the ambient medium is expanding with velocity HR_S at the radius R_S of the shock.

2. Various Self-Similar Solutions for Blast Waves

The classic problem of the evolution of an an adiabatic blast wave in a medium of constant density, was solved by Sedov and Taylor (see, for example, Sedov 1959). The radius R_S of the shock grows as power of time, which may be determined by noting that the kinetic energy of material swept up by the shock is comparable to the energy E released: $\rho R_S^3 \cdot (R_S/t)^2 \sim E$, or

$$R_S(t) = \kappa \left(\frac{Et^2}{\rho} \right)^{1/5} \propto t^{2/5}, \qquad (10)$$

where the constant $\kappa \sim 1$ is yet to be determined. Here, ρ is the ambient density. This result is valid only so long as the thermal energy of the material swept up by the blast wave is small, $\rho \cdot (kT/\mu) \cdot R_S^3 \ll E$; or, equivalently, so long as the shock is strong, $V_S^2 \gg \gamma kT/\mu = c^2$. Here, T is the temperature of the surrounding medium, which we assume to be gaseous, and c^2 is the adiabatic speed of sound. When this condition no longer holds, the blast wave weakens into an expanding sound wave.

It is instructive to consider this prototypical example in some detail. In writing down Eq. (10), we have assumed that the evolution of the blast wave is *self-similar*, in the sense that the quantities ρ, P, u, T may all be expressed as functions of r and t in the form $x(r,t) = X(t)\, \tilde{x}(r/R_S(t))$. In fact, this equation is accurate only when there is no characteristic timescale in the problem, such as may be introduced by effects like cooling, drag or, as just mentioned, the accumulation of thermal energy from the surrounding medium. But if such processes may be neglected, then, as has been shown by numerical simulations, a wide range of initial conditions converge to the self-similar solution, which is stable to both radial and non-radial perturbations.

In sum, we always find that the radius of a self-similar blast wave scales as

$$R_S(t) = R_S(t_0) \left(\frac{t}{t_0} \right)^{\eta}. \qquad (11)$$

The velocity of the shell with respect to the center of the blast is $U_S = \eta R_S/t$. In the case of a cosmological blast wave in an Einstein–de Sitter universe, the peculiar velocity

$$V_S = U_S - HR_S = \left(\frac{3}{2}\eta - 1 \right) HR_S. \qquad (12)$$

One direct consequence of self-similarity is that the kinetic and thermal energy (and, in the case of a cosmological blast wave, the potential energy) are *constant* fractions of the total conserved energy E. In this, we will find an important distinction between blast waves in a static medium, and blast waves in an expanding universe.

Whereas most of the energy is thermal in the Sedov solution ($E_{th}/E = 0.70$), the energy of a cosmological blast is mainly kinetic and potential ($E_{th}/E = 0.03$). One may therefore view a cosmological blast as a solitary wave in the Hubble flow. This results in important qualitative changes in the evolution of a blast wave. Complete numerical treatments of the non-self-similar evolution of a cosmological blast wave (cf. Ikeuchi, Tomisaka and Ostriker 1983) show that a very wide range of initial conditions do lead to the self-similar state after several doubling times.

It should also be noticed that, in the presence of dissipative effects like thermal conduction and viscosity, blast waves are *not* damped. The shock is smoothed out (indeed, it cannot exist without a finite viscosity), but energy is not lost by the blast wave, because the shock moves faster than the speed of sound in the surrounding medium. The bulk kinetic energy lost to dissipation is added to the thermal energy of the blast, and is restored as the blast sweeps up new material from the surrounding medium. Now, the density is very small at the center of an adiabatic blast wave, and the temperature is very high, so one effect of thermal conduction is to raise the density and lower the temperature; cf. Sedov (1959). (The density vanishes and the temperature is infinite at the center, under the assumption of a perfect gas equation of state with constant γ; but this assumption breaks down at a finite radius.)

With these thoughts, let us turn to the problem of an adiabatic blast wave in an expanding universe. In this case, the ambient density is decreasing in time, as $\rho \propto t^{-2}$, so that

$$R_S(t) = \kappa' \left(\frac{GEt^4}{\Omega_b} \right)^{1/5} \propto t^{4/5}, \quad (\eta = 4/5). \tag{13}$$

The peculiar velocity of the shell is $V_S = \frac{1}{5} H R_S$. If $\Omega_b = 1$, then this solution holds so long as the thermal energy of swept up material is less than E. But if $\Omega_b < 1$, we must consider the dynamical effects of the dark matter. In passing through the shell of baryons, the energy of dark matter increases by an amount $GM_S/R_S = \frac{1}{2}\Omega_b(HR_S)^2$ per unit mass. Therefore, the self-similar solution holds only so long as $\frac{1}{2}M_S(HR_S)^2 \ll E$. The details of what happens when this inequality ceases to be true are complicated. In short, if the dark matter is pressureless, then all the dark matter interior to the shell eventually passes the shell and oscillates about it. The asymptotic form of $R_S(t)$ is given approximately by the substitution $\Omega_b \to 1$ in Eq. (13).

Another important situation to consider is one where the material heated behind the shock can cool efficiently. This material is swept up in a very thin shell, and, when gravity may be neglected, the *momentum* of the shell $P_S = M_S U_S$ is conserved. Now, P_S may or may not be easy to determine, as follows.

In the case where a certain amount of mass ΔM is injected with kinetic energy E (as in a supernova explosion), both energy and momentum are conserved until a time t_M, when the mass swept up becomes comparable to ΔM. If the cooling time $t_{cool} < t_M$, then momentum is conserved throughout the expansion of the blast wave. But if $t_{cool} > t_M$, then the blast wave undergoes a period of adiabatic expansion with constantly increasing momentum, before cooling sets in. The subsequent evolution of the blast wave depends on the details of this intermediate phase.

Assuming that P_S has been determined, the radius of the shock scales as $\rho R_S^3(R_S/t) \sim P_S$, or

$$R_S(t) = \left(\frac{3P_S t}{\pi \rho} \right)^{1/4} \propto t^{1/4}, \quad (\eta = 1/4), \tag{14}$$

when the ambient medium has constant density. This is the Oort (1951) "snowplow" solution.

A related case is a hot bubble driving a cool shell; that is, the interior of the bubble is so rarefied that it cannot cool, but the denser material in the shell can cool. In this case, P_S does increase, but the blast wave steadily loses energy as the heated post-shock gas cools. To determine the motion of the shell, we must do a little more work. The equation of motion is

$$\frac{d(M_S U_S)}{dt} = 4\pi R_S^2 \cdot P_{int}, \tag{15}$$

and the pressure P_{int} of the hot interior evolves according to

$$P_{int}(t) = P_{int}(0) \left[\frac{R_S(t)}{R_S(0)} \right]^{-3\gamma}. \tag{16}$$

Equation (15) has the self-similar solution

$$R_S(t) = R_S(0) \left[\frac{3(2+3\gamma)^2}{6(2-\gamma)} \frac{P_{int}(0)}{R_S(0)^2} t^2 \right]^{2/(2+3\gamma)}. \tag{17}$$

So for $\gamma = \frac{5}{3}$, the radius of the shell scales as $R_s \propto t^{2/7}$ (Ostriker and McKee 1988), a dependence only slightly different from the case of pure momentum conservation.

We defer to later our discussion of a cosmological solution for a cold, radiative shell driven by a hot bubble.

Now, the present intergalactic medium is so rarefied that it never cools if heated at late epochs to temperatures $\gg 10^5 \, °\text{K}$. Nonetheless, at $z > 10$, hot electrons cool efficiently, by inverse Compton scattering off the cosmic background radiation. Once plasma has cooled and condensed sufficiently, other cooling processes such as free-free emission can become important. Recall that the momentum of a particle, measured in a comoving frame, decays in an expanding universe: $d\mathbf{p}/dt = -H\mathbf{p}$. The peculiar velocity of a cold, expanding shell decays analogously, and in a few expansion times it essentially comoves with with the Hubble flow. In the limit $\Omega_b \ll 1 = \Omega$, the mass of the shell may be neglected with respect to the mass interior to it, and

$$R_S(t)\left(\frac{t}{t_i}\right)^{-2/3} = R_S(t_i) + 3t_i \cdot V_S(t_i)\left[1 - \left(\frac{t}{t_i}\right)^{-1/3}\right]. \qquad (18)$$

If the initial peculiar velocity is $V_S(t_i) = f.HR_S(t_i)$, then the comoving radius increases asymptotically to only $(1 + 2f)$ times its initial value in an Einstein–de Sitter universe.

Next let us consider the evolution of a blast wave into which energy is *continuously* released — perhaps from an active galaxy, a quasar, or a superconducting string. In particular, we assume that the injection rate $\dot{E}$ is constant. Then the analog of the adiabatic Sedov solution is

$$R_S(t) = \kappa''\left(\frac{\dot{E}t^3}{\rho}\right)^{1/5} \propto t^{3/5}, \quad (\eta = 3/5), \qquad (19)$$

when the ambient medium has constant density. In an expanding universe, the relation is instead

$$R_S(t) = \kappa'''\left(\frac{G\dot{E}}{\Omega_b}\right)^{1/5} \cdot t \equiv \kappa'''V_E t \ (\eta = 1). \qquad (20)$$

The shock expands with a *constant* velocity $U_S = \kappa'''V_E$ and peculiar velocity $V_S = \frac{1}{3}U_S$. Shortly, we will discuss the case where the shock-heated gas may efficiently cool, but the hot interior of the blast wave suffers only adiabatic losses. Then the expression for R_S is identical to Eq. (20), except for a slight difference in the coefficient κ'''. After the energy release stops, the blast wave asymptotically approaches a comoving state.

Until now, we have been considering the effects of the release of energy in a small volume. Of course, it is possible an expanding shock will trigger additional heating, giving rise to a *detonation* wave. For example, the shocked gas may be sufficiently heated to undergo nuclear burning; alternately, is may cool sufficiently to be turned into stars, some of which explode as supernovae. The simplest prescription to take is that a fraction ε of the rest energy of the baryons swept up by the detonation wave is released as heat; in this case $M_S(R_S/t)^2 \sim \varepsilon M_S c^2$, and we find

$$R_S(t) \sim \varepsilon^{1/2}ct. \qquad (21)$$

The form of this expression does not depend on whether the ambient medium is static or time-dependent.

The complete solution for a detonation wave usually results in the following condition, originally postulated by Chapman and Jouget: in the rest frame of the shock, the motion of the post-shock gas is transonic, so that $u_2 = c_2$. This condition is sufficient to determine the motion of the wave,

$$R_S(t) = \left[2(\gamma^2 - 1)\varepsilon\right]^{1/2} ct, \tag{22}$$

a result which is compatible with our previous estimate (21). In the absence of burning, the Chapman–Jouget condition would imply that the shock was weak — in effect, a sound wave. But in this case, the shock is not weak. For example, the density increases behind the shock,

$$\frac{\rho_2}{\rho_1} = \frac{\gamma + 1}{\gamma}. \tag{23}$$

When the heating is by supernovae, we may estimate the velocity of the detonation wave by assuming that the energy released by the supernovae is that required to explain observed metallicities. Supposing that each supernova releases $3.10^{51}\,\mathrm{erg\,s^{-1}}$ of kinetic energy for every solar mass of metals, the heating rate per unit mass which results in a metallicity Z is $\varepsilon = 1.7 \times 10^{-3}Z$. For solar abundances, this is $\varepsilon = 3 \times 10^{-5}$, and the velocity of the shell is quite large, $1 \times 10^{-2} c = 3000\,\mathrm{km\,s^{-1}}$. This corresponds, for a cosmological blast wave, to a peculiar velocity of $1000\,\mathrm{km\,s^{-1}}$. However, this is almost certainly an overestimate. Cosmological blast waves fueled by nuclear energy are too small to explain directly the voids observed in in the distribution of galaxies, which have a typical radius of $1500\,\mathrm{km\,s^{-1}}$ (de Lapparent et al. 1986).

3. Shell Structure

A characteristic of all blast waves is that the material swept up is concentrated in a thin shell behind the shock. The simple reason for this is that the post-shock density is larger than that of the ambient medium. If the density in the shell were constant, then its thickness would be

$$\Delta R = \frac{1}{3}\frac{\gamma - 1}{\gamma + 1} R_S$$

$$= \frac{1}{12} R_S, \quad (\gamma = 5/3). \tag{24}$$

This is a reasonable estimate for a Sedov-Taylor blast wave.

In this respect, there is an important difference in behavior between an adiabatic blast wave in a homogeneous medium, and in an expanding universe (or a medium with density strongly decreasing from the center of the blast wave). In the former case, the density in the shell *decreases* monotonically from the shock to the center of the blast wave. In the latter case, the density increases from its postshock value to a very large value at a finite radius — which is the inner edge of the shell. This difference may be easily explained. The farther a fluid element lies from the shock, the earlier it passed through the shock. Denote by P^*, ρ^* and S^* the post-shock pressure, density and entropy at time t^*. In the approximation where P varies slowly through the shell, the density increases inward from the shock, if the entropy decreases; and *vice versa*. But the entropy of any fluid element is conserved after it passes through the shock, so we are lead to ask how the post-shock entropy scales with time. When the fluid is a perfect gas, the entropy $e^{S_*(\gamma-1)} \propto P_*/\rho_*^\gamma$. Now, in an expanding universe,

$$\rho^* \propto (t^*)^{-2} \quad \text{and} \quad P^* \propto \rho^* V_S^2(t^*) \propto (t^*)^{2\eta-4}.$$

Therefore $e^{S_*(\gamma-1)} \propto (t^*)^{2\gamma+2\eta-4}$, and the density increases inward from the shock if $\gamma > 2 - \eta$. We expect this condition to be satisfied since, for a point blast, it is $\gamma > \frac{6}{5}$; and when the rate of energy injection is constant, it is $\gamma > 1$.

We cannot take the limit $t^* \to 0$, which would imply an infinite density at the inner edge of the shell, since the blast wave presumably originates at some finite time, t_0. At times $t - t_0 \ll t_0$, the blast wave grows in a constant density medium, and the evolution of the blast follows the Sedov solution. That is, we expect ρ to reach some large but finite value at the inner edge of the shell, and then decrease smoothly to zero at the center of the blast wave (cf. Ikeuchi, Tomisaka and Ostriker 1983). Correspondingly, the pressure vanishes at the inner edge of the shell, and everywhere in the interior, only in the self-similar limit.

When the density of the ambient medium decreases from the center of the blast wave as $\rho \propto r^{-k_\rho}$, we may determine the structure of the shell in a similar manner. Using the self-similar solution (10), one finds that the density increases inward from the shock if (cf. Sedov 1959)

$$k_\rho > \frac{7 - \gamma}{\gamma + 1}$$
$$= 2 \quad (\gamma = 5/3). \tag{25}$$

We may conclude from this discussion that the shell of a cosmological blast wave is thinner than that of a blast wave in a static, homogeneous medium. For example, the adiabatic cosmological blast has $\Delta R = 0.033 \, R_S$ for $\gamma = \frac{5}{3}$ (Bertschinger 1983). The main reason is that the entropy gradient behind the shock is reversed, so that the density increases from the shock. Alternately, one may view the thinness of these shells as a direct corollary of the small thermal energy of the blast. Indeed, a cooling shell is even thinner; its thickness is determined by the equilibrium temperature of the shell and the requirement of pressure balance.

4. Equation of Motion of a Thin Shell

Under the assumption that the mass swept up by the blast wave is concentrated in a thin shell (of thickness ΔR), we can write down a simple equation of motion for the shell. Taking the integral $\int_{R_S-\Delta R}^{R_S} dr 4\pi r^2 \rho$ of both sides of Euler's equation (2), we obtain

$$\frac{\partial}{\partial t}(M_S U_S) = P_{\text{int}} \cdot 4\pi R_S^2 - \frac{1}{2} M_S H^2 R_S \left(1 - \frac{1}{2}\Omega_b\right)$$

$$+ H R_S \frac{\partial M_S}{\partial t} + M_S f(R_S), \tag{26}$$

where $P_{\text{int}} = P(R_S - \Delta R)$ is the pressure at the inner edge of the shell. We have assumed a strong shock and neglected the pressure of the ambient medium. The mean gravitational acceleration through the shell is $-\frac{1}{2}H^2 R_S \left(1 - \frac{1}{2}\Omega_b\right)$, and the term $H R_S(\partial M_S/\partial t)$ represents the momentum collected by the shell from the Hubble flow.

In an adiabatic blast wave, one cannot calculate P_{int} accurately without introducing further approximations (or calculating the full self-similar solution). The reason is that the decrease in the thermal energy of the hot interior due to expansion is balanced by heating behind the shock, and this second effect is difficult to account for. The result of an exact calculation for the undetermined constant κ in the Sedov solution, Eq. (10), is (Sedov 1959),

$$R_S(t) = 1.152 \left(\frac{Et^2}{\rho}\right)^{1/5} \quad (\gamma = 5/3; \quad \rho = \text{constant}),$$

$$R_S(t) = 1.033 \left(\frac{Et^2}{\rho}\right)^{1/5} \quad (\gamma = 7/5; \quad \rho = \text{constant}). \tag{27}$$

It is now possible to estimate the relative contributions of the kinetic, thermal and potential energy to the total energy of a blast wave. In the thin-shell approximation, we take the velocity of the shell to be the *post-shock* velocity, and the kinetic energy is

$$E_k = \frac{1}{2} M_S \left(\frac{2U_S}{\gamma+1}\right)^2 = \frac{2\eta^2}{(\gamma+1)^2} M_S \left(\frac{R_S}{t}\right)^2 \tag{28}$$

for a blast wave in a static medium, and

$$E_k = \frac{1}{2} M_S \left(\frac{2U_S}{\gamma+1} + \frac{\gamma-1}{\gamma+1} H R_S\right)^2 - \frac{2}{15} M_S \left(\frac{R_S}{t}\right)^2$$

$$= \left[\frac{1}{2}\left(\frac{6\eta + 2\gamma - 2}{3(\gamma+1)}\right)^2 - \frac{2}{15}\right] M_S \left(\frac{R_S}{t}\right)^2 \tag{29}$$

for a cosmological blast wave (we have subtracted the kinetic energy of the unperturbed Hubble flow). We assume a self-similar solution, $R_S \propto t^\eta$.

The potential energy of a cosmological blast wave is (relative to the Hubble flow)

$$E_\phi = -\frac{GM_S^2}{2R_S} + \frac{2}{15} M_S \left(\frac{R_S}{t}\right)^2 = \frac{1}{45} M_S \left(\frac{R_S}{t}\right)^2 \tag{30}$$

when $\Omega_b = 1$. Thus, E_ϕ is *positive*. The change in the potential energy relative to the Hubble flow is small compared to the change in the kinetic energy for the most important self-similar solutions which we have catalogued above. The potential energy is also small when $\Omega_b \ll 1$. We ignore the effects of gravity if the ambient medium is static.

For the Sedov solution, we have $M_S(R_S/t)^2 = \frac{4\pi}{3}\kappa^5 E$. In the thin shell approximation, we then have $E_k/E = 0.38$ ($\gamma = \frac{5}{3}$). The thermal energy is $E_{th}/E = 0.62$, and is the dominant contribution to the total energy. (The exact self-similar solution of the Sedov-Taylor problem gives $E_k/E = 0.30$ and $E_{th}/E = 0.70$; we will find that the thin-shell approximation works much better in a cosmological blast wave.)

In the approximation where the pressure is constant through the shell, the thermal energy of a cosmological blast wave very small,

$$
E_{th} = \frac{2}{\gamma - 1} M_S \frac{kT_{ps}}{\mu}
$$

$$
= \frac{2}{(\gamma + 1)^2}\left(\eta - \frac{2}{3}\right)^2 M_S \left(\frac{R_S}{t}\right)^2. \tag{31}
$$

Here, T_{ps} is the post-shock temperature. Such an estimate is not accurate for the Sedov solution, because P changes only by a factor of two or three between the center and the shock. The thermal energy of the blast wave is dominated by the very hot and rarefied interior of the blast wave. On the other hand, the pressure goes to zero at the inner edge of the shell in the cosmological analog of the Sedov solution. The thermal energy of the blast wave is contained entirely by the shell.

Let us now compare the various contributions to the total energy of an adiabatic cosmological blast wave. Substituting $\eta = \frac{4}{5}$ in Eqs. (29)–(31) and taking $E_k + E_\phi + E_{th} = E$, we find $E_k/E = 0.85$, $E_\phi/E = 0.12$ and $E_{th}/E = 0.03$ (for $\gamma = \frac{5}{3}$). These results are in excellent agreement with those obtained from the exact self-similar solution (Bertschinger 1983; Ikeuchi, Tomisaka and Ostriker 1983). We may also solve for the unknown constant κ' in Eq. (13), obtaining

$$
R_S(t) = 1.887 \left(\frac{GEt^4}{\Omega_b}\right)^{1/5} \quad (\gamma = 5/3;\ \rho \propto t^{-2}),
$$

$$
R_S(t) = 1.869 \left(\frac{GEt^4}{\Omega_b}\right)^{1/5} \quad (\gamma = 7/5;\ \rho \propto t^{-2}) \tag{32}
$$

The difference between these two values is small because of the relative unimportance of the thermal energy. Since the thermal energy of a cosmological blast wave is small, the blast wave is essentially a rearrangement of the Hubble flow.

The thin-shell approximation is sufficient to calculate the motion of a blast wave, in the case where the hot interior of the blast wave cannot cool, but the postshock gas can. First let us consider a point blast wave. The pressure P_{int} of the hot interior decreases according to Eq. (16). We look for a self-similar solution of Eq. (26), and obtain in the case $\Omega_b = 1$ (Ostriker and McKee 1988),

$$
R_S(t) \propto t^{(15 + \sqrt{17})/24} = t^{0.797}. \tag{33}
$$

The energy decays as

$$E \propto t^{-(21-5\sqrt{17})/24} = t^{-0.016}.$$ (34)

That is, the introduction of post-shock cooling results in only a very slight change from the adiabatic self-similar solution, which was $R_S \propto t^{0.8}$. The main reason for this is, of course, that the thermal energy of the blast wave is small. The net energy of the blast decreases very slowly, since the losses during each expansion by a factor of two are only of order E_{th}. The various contributions to the total instantaneous energy of the blast wave are $E_k/E = 0.88$, $E_\phi/E = 0.12$ and $E_{th}/E = 0$ (in the limit of zero cooling time).

A cooling shell driven by a hot bubble is the relevant problem when the active ingredient is a superconducting cosmic string (Ostriker, Thompson and Witten 1986). The hot interior may either be filled by ultra-low frequency electromagnetic radiation, or a hot plasma which is so rarefied that electrons and ions do not couple. The pressure is then constant throughout the interior, and evolves according to the first law of thermodynamics,

$$\frac{\partial P_{\text{int}}}{\partial t} = \frac{\dot{E}}{4\pi R_S^3} - 4P\frac{U_S}{R_S},$$ (35)

when the interior is filled with radiation. The first term on the RHS represents heating by the string, and the second adiabatic losses due to the expansion of the bubble. We know already that $U_S = R_S/t$, so the solution to Eq. (35) is

$$P_{\text{int}} = \frac{\dot{E}t}{8\pi R_S^3}.$$ (36)

Substituting this expression in the equation of motion (26), and assuming the self-similar solution, we find for the coefficient κ''' in Eq. (20):

$$R_S(t) = \left(\frac{81}{20}\frac{G\dot{E}}{\Omega_b}\right)^{1/5} \cdot t = 1.323\, V_E t,$$ (37)

where V_E is defined in Eq. (20). If instead the interior of the bubble is filled with a hot plasma, and non-relativistic ions provide the main contribution to P_{int}, then the result is

$$R_S(t) = \left(\frac{27}{10}\frac{G\dot{E}}{\Omega_b}\right)^{1/5} \cdot t = 1.220\, V_E t.$$ (37')

5. Gravitational Instability in a Shell

Cosmological blast waves are subject to a number of instabilities. The simplest instability to consider is that induced by the self-gravity of the expanding shell. Now, an adiabatic blast wave is stable even in the presence of gravity. The basic reason for this is that the sound speed is high enough in the blast for pressure to balance the attractive force of gravity at all length-scales. When the blast is able to cool, however, the density of the shell may be high enough for the shell to be gravitationally unstable. We may estimate the characteristic size of the fragments produced by this instability, as follows (Ostriker and Cowie, 1981).

Consider a circular patch of a smooth shell with surface mass density Σ. The patch has radius a, and the shell has radius R_S. In a frame comoving with the shell, the energy per unit mass of the patch is

$$E = E_k + E_\phi + E_{th} = \frac{a^2}{4}\Sigma\left(\frac{U_S}{R_S}\right)^2 - k\pi a G\Sigma + e. \tag{38}$$

Here, we have assumed that the shell is isothermal, with internal energy e per unit mass. k is a numerical coefficient of order unity. For a uniform, isolated disk, $k = 0.849$. The kinetic term appears because the shell is expanding, so that an observer comoving with the shell experiences an effective two-dimensional Hubble expansion. The patch is most strongly bound at the value of $a = a_{\rm crit}$ where $dE/da = 0$. This is

$$\frac{a_{\rm crit}}{R_S} = 2\pi k\frac{G\Sigma R_S}{U_S^2}. \tag{39}$$

The patch is bound ($E < 0$ at $a = a_{\rm crit}$) only for

$$e < \left(\frac{k\Omega_b}{18\eta}\right)^2\left(\frac{R_S}{t}\right)^2. \tag{40}$$

This means that a shell of comoving radius R_S^c is gravitationally unstable only if its temperature is less than

$$T < 4.10^3 \,^\circ\text{K}\,\frac{k}{\eta}(1+z)\left(\frac{\Omega_b}{0.1}\right)\left(\frac{R_S^c}{1500\,\text{km}\,\text{s}^{-1}}\right). \tag{41}$$

Here, we have taken $e = 3kT/m_p$.

The characteristic mass of the shell fragments is $\Delta M = \pi a_{\rm crit}^2\Sigma$, or

$$\Delta M = \frac{k^2}{324\eta^4}\Omega_b^2 M_S = 2\cdot 10^{10} M_\odot\frac{k^2}{\eta^4 h_{50}}\left(\frac{\Omega_b}{0.1}\right)^3\left(\frac{R_S^c}{1500\,\text{km}\,\text{s}^{-1}}\right)^3. \tag{42}$$

Note that ΔM does not depend on the temperature of the shell, in so far as the shell is unstable. The above analysis is confirmed by detailed numerical simulations (White and Ostriker 1988) of a collisionless gas shell.

We should add that, irrespective of whether it is gravitationally unstable, a cooling shell *does* suffer from Rayleigh–Taylor instabilities (Vishniac 1983).

6. Interactions between Cosmological Blasts

Finally, let us consider briefly what may happen when cosmological blast waves interact, and how these interactions may be changed by the presence of dark matter.

There are various ways that one could determine that isolated cosmological blast waves were comprised of baryons moving through a more uniform background of dark matter. All of these would be difficult to implement, in the absence of a knowledge of the original blast energy, or current peculiar velocity of the shells.

However, the interactions between shells are substantially different when all, or only a fraction, of matter is swept up in the blast waves. For example, if the fraction of closure density in baryons were $\Omega_b = 1$, then numerical simulations (Dekel, Ostriker and Weinberg 1987) show that the common membrane between two interacting shells will disappear rapidly. After considerable overlap occurs, most of the matter will drain to the common vertices (cf. the numerical simulations of Peebles 1987), so that the shells will be replaced by linear and point-like features. But if $\Omega_b \ll 1$, then collisionless shells of galaxies will pass through each other largely unaffected by their mutual gravity. It is easy to show that the mass which accumulates at the places where three shells intersect is only a fraction $\sim \Omega_b^2$ of the mass in each of the shells. Preliminary work indicates (Dekel, Ostriker and Weinberg 1987) that the fraction of the baryonic mass accumulated in vertices is comparable to the fraction of galaxies in rich Abell clusters.

References

Bertschinger, E. W. (1983) *Astrophys. J.* **268**, 17.

de Lapparent, V., Geller, M. J., and Huchra, J. P. (1986) *Astrophys. J. Lett.* **302**, L1.

Dekel, A., Ostriker, J. P., and Weinberg, D. (1987) In preparation.

Guth, A. (1981) *Phys. Rev.* **D23**, 347.

Ikeuchi, S., Tomisaka, K., and Ostriker J. P. (1983) *Astrophys. J.* **265**, 583.

Landau, L. D. and Lifshitz, E. M. (1959) *Fluid Mechanics*, Pergamon, London, U.K.

Loh, E. D. and Spillar, E. J. (1986) *Astrophys. J. Lett.* **307**, L1.

Oort, J. H. (1951) In: *Problems of Cosmical Aerodynamics*, p. 118, Central Air Documents Office, Dayton OH, U.S.A.

Ostriker, J. P. and Cowie, L. L. (1981) *Astrophys. J. Lett.* **243**, L127.

Ostriker, J. P. and McKee, C. M. (1988) *Rev. Mod. Phys.* in press.

Ostriker, J. P., Thompson, C., and Witten, E. (1986) *Phys. Lett.* **B280**, 231.

Peebles, P. J. E. (1987) *Astrophys. J.* **317**, 576.

Sedov, L. I. (1959) *Similarity and Dimensional Methods in Mechanics*, Academic Press, New York, U.S.A.

Vishniac, E. (1983) *Astrophys. J.* **274**, 152.

Chapter 6

DARK MATTER IN GALAXIES AND GALAXY SYSTEMS

Scott Tremaine* and Hyung Mok Lee[†]

Canadian Institute for Theoretical Astrophysics,
University of Toronto, Toronto, Canada

We review the dynamical evidence for dark matter in galaxies, groups of galaxies and clusters of galaxies. A summary, expressed in terms of the mass-to-light ratios of various systems, is presented in Fig. 3 near the end of the review.

1. Introduction

In astronomy, most information regarding the presence of different kinds of mass comes from photons at various wavelengths. Very hot gases emit X-rays, while stars produce most of their energy at optical wavelengths. Some atomic or molecular gases in interstellar space show emission lines at radio wavelengths (e.g. HI, CO, etc.). In addition, there are non-luminous objects whose existence can be inferred from other considerations. For example, interstellar dust grains are known to exist because of interstellar reddening, and the number of non-luminous stellar remnants (e.g. black holes, neutron stars or white dwarfs) can often be estimated from stellar population and stellar evolution theory.

Sometimes, certain masses manifest themselves only through gravitational interaction. We will use the term *dark matter* (abbreviated as DM) to denote matter whose existence is inferred only through its gravitational effects. Therefore the best — indeed only — way of studying dark matter is to accurately determine the mass of astronomical objects from their dynamics and to compare this mass with the mass inferred from the light emitted by the objects. A discrepancy indicates the presence of dark matter.

The determination of dynamical mass, however, is not a trivial task. The main difficulty is that even perfect observations (i.e. observations without error) cannot always provide enough information to completely constrain theoretical models. This is because we always observe projected positions on the plane of the sky and line-of-sight velocities at a given instant rather than complete three-dimensional orbits over an extended period of time. The subject of these lectures is the methods of determining dynamical masses of galaxies and systems of galaxies. The plan of the lectures is as follows: the basic theoretical framework is discussed in the next lecture.

---

*Current address: Princeton University Observatory, Princeton, NJ 08544, USA.
[†]Current address: Department of Astronomy, Seoul National University, Seoul 151-742, Korea.

The subsequent four lectures describe the determination of masses of various systems. The final lecture provides a summary. Additional details and other topics are provided in many review articles, conference proceedings, and textbooks, including Binney and Tremaine (1987), Faber and Gallagher (1979), Kormendy and Knapp (1987), Primack (1987), and Trimble (1987).

We would like first to discuss some introductory subjects that do not fit easily into the main arguments in the other lectures.

1.1. *Virial Theorem*

One of the simplest ways to determine the mass of a stellar system is through the virial theorem.

Consider a self-gravitating system composed of N point masses. We denote by m_i, $\vec{r}_i$, and $\vec{v}_i$ the mass, position and velocity of each particle. The moment of inertia of such a system is

$$I = \sum_{i=1}^{N} m_i \vec{r}_i^2. \tag{1.1}$$

Now take first and second time derivatives of the moment of inertia to get

$$\dot{I} = 2 \sum_{i=1}^{N} m_i \vec{r}_i \cdot \vec{v}_i \tag{1.2}$$

and

$$\ddot{I} = 2 \sum_{i=1}^{N} m_i (\vec{v}_i^{\,2} + \vec{r}_i \cdot \ddot{\vec{r}}_i). \tag{1.3}$$

Since the system is self-gravitating, the acceleration of particle i can be computed by summing over the contribution from all the particles

$$\ddot{\vec{r}}_i = \sum_{j \neq i} G m_j \frac{(\vec{r}_j - \vec{r}_i)}{|\vec{r}_j - \vec{r}_i|^3}, \tag{1.4}$$

so that Eq. (1.3) can be rewritten as

$$\frac{1}{2}\ddot{I} = 2K + \sum_{i=1}^{N}\sum_{j \neq i} G m_i m_j \frac{\vec{r}_i \cdot (\vec{r}_j - \vec{r}_i)}{|\vec{r}_j - \vec{r}_i|^3}$$

$$= 2K - \frac{G}{2}\sum_{i=1}^{N}\sum_{j \neq i} \frac{m_i m_j}{|\vec{r}_i - \vec{r}_j|} = 2K + W, \tag{1.5}$$

where

$$K = \frac{1}{2}\sum_{i=1}^{N} m_i \vec{v}_i^{\,2}, \tag{1.6}$$

is the total kinetic energy and

$$W = -\frac{G}{2} \sum_{i=1}^{N} \sum_{j \neq i} \frac{m_i m_j}{|\vec{r}_i - \vec{r}_j|}, \tag{1.7}$$

is the total gravitational potential energy of the system.

If the system is in a stationary state, we may assume that $\langle \ddot{I} \rangle_t = 0$, where $\langle \rangle_t$ is the time average taken over several times the dynamical timescale of the system. The dynamical timescale is the time required for typical particles to complete one orbit, that is, to move across the whole system. In practice, we observe at only one epoch. Thus, although the rigorous form of the virial theorem is $2\langle K \rangle_t + \langle W \rangle_t = 0$, in practice the theorem is used without the time average as follows:

$$2K + W \approx 0. \tag{1.8}$$

Equation (1.8) is valid only if the system is in equilibrium and in a stationary state. Such a condition is achieved if the age of the system is much longer than its dynamical timescale.

The virial theorem formed the basis for most mass determinations of galaxies and galaxy systems for decades, and although it has now largely been superseded by more sophisticated methods, it is still used to provide rough mass estimates for systems such as groups of galaxies with few members where detailed modelling is inappropriate.

1.2. *History of Dark Matter*

The discovery of dark matter can be clearly traced to a seminal paper by Zwicky (1933). By the early 1930's, Hubble's law relating radial velocities v of distant galaxies to their distances d through

$$v = H_0 d, \tag{1.9}$$

was established, although the Hubble constant H_0 was very poorly determined. The above relation holds for galaxies with $v \gtrsim 2000 \, \mathrm{km \, s^{-1}}$. We will parametrize the Hubble constant as

$$H_0 = 100 \, h \, \mathrm{km \, s^{-1} \, Mpc^{-1}}, \tag{1.10}$$

where $1 \, \mathrm{Mpc} = 1$ megaparsec $= 3.086 \times 10^{24} \, \mathrm{cm}$ and h is a dimensionless parameter. In the early 1930's, h was known to be 5.58, while it is currently believed to lie between 0.5 and 1.

At the time of Zwicky's paper, rotation curves were available for some spiral galaxies so that the mass-to-light ratios of galaxies were determined to be

$$\frac{M}{L} \approx 1 \, h \left(\frac{\mathrm{M}_\odot}{\mathrm{L}_\odot} \right)_V. \tag{1.11}$$

The subscript V in the above equation indicates that the luminosity is measured in the visual band. The unit for mass-to-light ratios adopted in the above equation, i.e. $(M_\odot/L_\odot)_V$, will be used throughout these lectures, and we will usually omit the units in expressing M/L from here on.

For $h = 5.58$, the mass-to-light ratio in Eq. (1.11) was consistent with observations of the Solar neighbourhood, which gave $M/L \approx 3$, not far from the Solar value. This agreement, together with the fact that galaxy spectra resembled the Solar spectrum, led to the simple and compelling picture that galaxies are made up of stars like the ones around us and that the Sun is a typical star in a galaxy.

Zwicky applied the virial theorem to the Coma cluster of galaxies. Radial velocities were known for seven galaxies in Coma at that time. The mean and root mean square velocities Zwicky used were

$$\overline{v_\parallel} \approx 7300\,\text{km}\,\text{s}^{-1}; \quad v_{\parallel,\text{rms}} = \left[\frac{1}{N-1}\sum_{i=1}^{N}(v_{\parallel,i} - \overline{v_\parallel})^2\right]^{1/2} \approx 700\,\text{km}\,\text{s}^{-1}. \qquad (1.12)$$

For spherical systems, the gravitational potential energy can be expressed in the form

$$W = -\alpha\frac{GM^2}{R}, \qquad (1.13)$$

where M and R are the total mass and radius of the system, respectively, and α is a constant depending on the density distribution. For a uniform sphere $\alpha = 3/5$, while $\alpha = 3/(5-n)$ for a polytropic sphere with polytropic index n. Note that this constant is not very sensitive to the specific form of the density distribution, which was poorly known at that time; thus the value for the uniform sphere ($\alpha = 3/5$) was used by Zwicky. If the cluster is spherical and the galaxies have the same mass, the total kinetic energy can be computed from the formula $K = \frac{3}{2}Mv_{\parallel,\text{rms}}^2$. After plugging these numbers into Eq. (1.8), he obtained the total mass of the Coma cluster

$$M \approx \frac{5v_{\parallel,\text{rms}}^2 R}{G} \approx 1 \times 10^{15}\,h^{-1}\text{M}_\odot, \qquad (1.14)$$

where he took the size of the cluster to be $1.5° \approx 2h^{-1}$ Mpc. The total luminosity of the cluster was known to be

$$L_V \approx 2 \times 10^{13}h^{-2}\text{L}_\odot, \qquad (1.15)$$

which gives the mass-to-light ratio $M/L \approx 50\,h$ in Solar units. This is about a factor of 50 larger than the M/L's of individual galaxies [Eq. (1.11)], whatever the value of h may be. Therefore he concluded that a large amount of non-luminous matter is required if the Coma cluster is in dynamical equilibrium.

By modern standards, Zwicky's analysis has several problems, including crude estimates for the cluster radius, luminosity, and density distribution, poor statistics due to the small number of radial velocities, and possible contamination from background or foreground galaxies. Nevertheless his principal result has survived. The mass-to-light ratios for both galaxies and clusters of galaxies have gone up, but the discrepancy of a factor of 50 found by Zwicky still remains. More detailed analyses show that the best present estimate of M/L for the Coma cluster is about $400\,h$, as will be discussed in greater detail in lecture 6.

Zwicky's remarkable result implies that at least 95% of the mass in Coma is in some invisible form. It suggests that on scales larger than $\approx 1\,\text{Mpc}$, visible stars represent only a minor contaminant in a vast sea of dark matter of completely unknown nature.

1.3. A Quick Review of Cosmology

One of the most important assumptions in standard cosmological models is that the Universe is isotropic and homogeneous on large scales (say, between $30\,h^{-1}$ and $3000\,h^{-1}$ Mpc). This means that there exists a set of "fundamental observers" for whom the Universe looks isotropic, and a cosmic time such that Universe is homogeneous for all fundamental observers at any fixed time. Any observer living in a galaxy is a fundamental observer to a good approximation. Of course, this is not the only available cosmological model, but one of the simplest and the most widely used one, and the assumption of large-scale homogeneity and isotropy is consistent with observations of galaxy counts and the microwave background. [This part of the lecture is mainly based on Gunn (1978).]

Suppose $l_{ij}(t)$ denotes the distance between two fundamental observers i and j, which can be written as

$$l_{ij}(t) = l_{ij}(t_0)R(t), \tag{1.16}$$

where t_0 is the present time and $R(t)$ is the scale factor, which depends only on t because the Universe is isotropic and homogeneous. Note that $R(t_0) = 1$. The subscript 0 in the above equation represents the value at the present epoch. The relative velocity between i and j is simply the time derivative of l_{ij}:

$$v_{ij}(t) = l_{ij}(t_0)\dot{R}(t) = \frac{\dot{R}(t)}{R(t)}l_{ij}(t) = H(t)l_{ij}(t), \tag{1.17}$$

which is equivalent to Hubble's law of expansion with the Hubble constant being

$$H(t) = \frac{\dot{R}(t)}{R(t)}; \quad H_0 = H(t_0). \tag{1.18}$$

Next, consider a non-relativistic free particle passing a fundamental observer at point A with a velocity v_p at time t and passing an observer at point B at a later time $t + dt$. If the separation between the points A and B is dl, the relationship between dt and dl is

$$dl = v_p dt. \tag{1.19}$$

The velocity seen at the point B is

$$v'_p = v_p - H(t)dl, \tag{1.20}$$

so that

$$\frac{dv_p}{dt} = -H(t)v_p, \tag{1.21}$$

which can be integrated to give the relation $v_p \propto 1/R(t)$. For the case of relativistic particles like photons, a similar argument gives the relationship between the frequency at which the photon is emitted and the frequency observed by us,

$$\frac{\nu_0}{\nu_e} = \frac{R(t_e)}{R(t_0)} \equiv \frac{1}{1+z}, \tag{1.22}$$

where the subscript e represents the time of emission and z is the redshift of the photon.

Now let us consider dynamics. We will use Birkhoff's theorem of general relativity, which states that in a spherical system the acceleration at any radius depends only on the mass distribution within that radius. Thus if we consider two galaxies separated by a distance $l(t)$, when l is sufficiently small their relative acceleration is described by the Newtonian formula

$$\frac{d^2l}{dt^2} = -\frac{GM}{l^2}, \tag{1.23}$$

where M is the mass inside a sphere of radius l. If the Universe is matter-dominated, then M is constant in time [since the Hubble flow is smooth, galaxies neither enter nor leave the comoving sphere of radius $l(t)$], and Eq. (1.23) can be integrated over t to yield

$$\frac{1}{2}\dot{l}^2 - \frac{GM}{l} = E, \tag{1.24}$$

where

$$M = \frac{4}{3}\pi\rho(t)l^3; \quad l = l_0 R(t), \tag{1.25}$$

and E is the integration constant, which is equivalent to binding energy in Newtonian dynamics. Dividing through by $\frac{1}{2}l_0^2$, we get

$$\dot{R}^2 - \frac{8}{3}\pi G\rho R^2 = \epsilon, \tag{1.26}$$

where $\epsilon = 2E/l_0^2$. At the present epoch, Eq. (1.26) can be written

$$H_0^2 - \frac{8}{3}\pi G\rho_0 = \epsilon, \tag{1.27}$$

or

$$1 - \Omega_0 = \frac{\epsilon}{H_0^2}, \tag{1.28}$$

where the density parameter Ω_0 is defined to be the value at the present epoch of

$$\Omega = \frac{8\pi G\rho}{3H^2}. \tag{1.29}$$

We may also express the density parameter as

$$\Omega_0 = \rho_0/\rho_c, \tag{1.30}$$

where the critical density

$$\rho_c \equiv \frac{3H_0^2}{8\pi G} = 1.88 \times 10^{-29}\, h^2 \mathrm{g\,cm}^{-3} = 2.76 \times 10^{11}\, h^2\, \mathrm{M_\odot Mpc}^{-3}, \tag{1.31}$$

which corresponds the present mean density required to make the Universe bound ($\epsilon < 0$).

The solutions to Eq. (1.26) can be divided into three cases: $\Omega_0 < 1$, $\Omega_0 > 1$ and $\Omega_0 = 1$:

(1) $\Omega_0 < 1$: In this case the parametrized solution to Eq. (1.26) is

$$R = A(\cosh\eta - 1), \quad t = B(\sinh\eta - \eta), \tag{1.32}$$

where

$$\Omega_0 = \mathrm{sech}^2\left(\frac{1}{2}\eta_0\right), \quad \frac{A^3}{B^2} = \frac{4\pi G\rho_0}{3}. \tag{1.33}$$

This solution represents a model for an open Universe. The curvature of three-dimensional space for this solution is negative and the Universe expands forever.
(2) $\Omega_0 > 1$: The parametrized solution for this case is

$$R = A(1 - \cos\eta), \quad t = B(\eta - \sin\eta), \tag{1.34}$$

where

$$\Omega_0 = \sec^2\left(\frac{1}{2}\eta_0\right), \quad \frac{A^3}{B^2} = \frac{4\pi G\rho_0}{3}. \tag{1.35}$$

This solution represents a closed Universe which eventually recollapses after maximum expansion. The space part of this solution has positive curvature.
(3) $\Omega_0 = 1$: Finally, in this case, the solution for Eq. (1.26) becomes

$$R = \left(\frac{3H_0 t}{2}\right)^{2/3}, \tag{1.36}$$

which represents a flat Universe. This solution also expands forever, but the expansion rate becomes asymptotically zero. The space part of this solution has zero curvature, and can be represented by Euclidean geometry.

The determination of Ω_0 from observation is of great importance because this parameter determines the future evolutionary path of the Universe. Notice that the value of Ω changes through the evolution, but the sign of $\Omega - 1$ remains unchanged.

1.4. *Mass-to-light Ratio in the Solar Neighbourhood*

The most important indicator of the presence of dark matter is the mass-to-light ratio, M/L. It is very instructive to know M/L in the region close to the Sun, where high precision observations are possible. Since we have a reasonably good idea of the constituents in the Solar neighbourhood — certainly better than in any other system — comparison of the dynamically determined M/L with that deduced from an inventory of the constituents serves to indicate the amount of dark matter more accurately than in more distant systems.

Since different stellar populations have different scale heights, the surface density integrated along the direction perpendicular to the Galactic plane is a more fundamental quantity than the volume density. Table 1 summarizes the surface density of mass and luminosity per square parsec integrated within $|z| < 700$ pc, where z is the height above the Galactic midplane.

From this table we deduce M/L in the Solar neighbourhood to be $\Sigma/I \approx 3.3$ for all known constituents.

The total mass density can be determined from carefully selected star samples by analyzing the velocity dispersion and density profile in the direction normal to the Galactic plane. These studies give the mass-to-light ratio within 700 pc,

$$\frac{M}{L}(|z| < 700\,\text{pc}) \approx 5, \tag{1.37}$$

which is slightly larger than that given by Table 1. Therefore, there is marginal evidence for dark matter in the Solar neighbourhood. This subject is discussed in greater detail in the lectures by John Bahcall.

The M/L's derived for the Solar neighbourhood are only benchmarks and should not necessarily be expected to apply to other systems. For example, roughly 95% of the light in the Solar neighbourhood comes from stars brighter than the sun but about 75% of the mass is contained in the stars fainter than the sun. Therefore, any slight variation of the initial mass function can change M/L significantly.

1.5. *Classification Scheme of Dark Matter*

Dark matter has been discussed in various astronomical contexts, and it is worth bearing in mind that both the reliability of the evidence and the nature of the dark

Table 1. Surface density and brightness of Solar neighbourhood.

Species	$\Sigma(\text{M}_\odot/\text{pc}^2)$	$I(\text{L}_\odot/\text{pc}^2)$
Visible stars	27	15
Dead stars	18	0
Gas	5	0
Total	50	15

matter may be quite different in different contexts. There are at least four different categories of possible dark matter (DM):

(1) DM in the Solar neighbourhood,
(2) DM in galaxies,
(3) DM in clusters and groups of galaxies, and
(4) DM in cosmology.

As we discussed above, there is marginal evidence for DM in the Solar neighbourhood. This is based on the discrepancy between the mass determined by local vertical dynamics and the mass detected by direct observations. The study of the Solar neighbourhood is important even though not much DM may be present, because it is here that we have best hope of determining the nature of the DM. One further property of the DM in the Solar neighbourhood is that it is concentrated in a disk, and hence must almost surely be composed of baryons, since dissipation in a rotating gas is by far the most common way to form disks.

Stronger evidence for DM can be found in galaxies. Recent 21-cm radio observations have revealed the ubiquity of flat rotation curves in spiral galaxies out to radii much larger than the radii containing most of the visible stars. (It is much more difficult to measure the rotation curves for elliptical galaxies.) If the mass is proportional to the light, the rotation curves should exhibit a Keplerian falloff at large radii (that is, $V_{\rm rot} \propto r^{-1/2}$). The flat rotation curves, instead, suggest that the mass within the radius r

$$M(r) \approx \frac{V_{\rm rot}^2 r}{G} \propto r,\tag{1.38}$$

which yields M/L's of up to 30–40 for individual galaxies and possibly much more depending on the extent of the flat rotation curve beyond the last measured point. If the rotation curves remain flat to several hundred kiloparsecs (as proposed by Ostriker et al. 1974), the M/L's of individual spiral galaxies may be comparable to those in clusters.

Clusters of galaxies have provided the best evidence for DM ever since Zwicky's original work. The Coma cluster of galaxies provides perhaps the single strongest piece of evidence for DM because of the dramatic difference in its M/L from that of the Solar neighbourhood (see lecture 6).

The principal cosmological evidence for DM is that theoretical prejudice, as well as some specific models of the early Universe such as inflationary models, require $\Omega_0 = 1$. To determine what M/L this implies we must estimate the mean luminosity density. The determinations of the present average luminosity density in the Universe by Davis and Huchra (1982) and Kirshner et al. (1983) can be averaged to give

$$j_0 \approx 1.7 \times 10^8 \, h (\mathrm{L}_\odot \mathrm{Mpc}^{-3})_V.\tag{1.39}$$

Using Eq. (1.31), the density parameter can be expressed as

$$\Omega_0 = \frac{\rho_0}{\rho_c} = \frac{M}{L}\frac{j_0}{\rho_c} \approx 6.1 \times 10^{-4} h^{-1} \left(\frac{M}{L}\right)_V \approx \frac{(M/L)_V}{1600\,h}, \qquad (1.40)$$

which means that M/L should be about $1600\,h$ in order for Ω_0 to be unity. This M/L is larger than that of the Coma cluster by about a factor of four. There is no known system of galaxies whose dynamical mass implies a mass-to-light ratio as high as $1600\,h$; hence, if the Universe is closed, the mass-to-light ratio must be much larger outside galaxy systems — even those as large as Coma — than within them.

An independent constraint on the total amount of baryonic matter in the Universe comes from the study of primordial nucleosynthesis. The present amount of deuterium and helium is mainly produced during an early phase of the Universe. The observed abundances of deuterium and helium set a limit on Ω_0 in the form of baryonic matter

$$\Omega_B \approx (0.011 - 0.048)h^{-2}, \qquad (1.41)$$

which translates to $20\,h^{-1} < (M_B/L)_V < 80\,h^{-1}$ through Eq. (1.40), where M_B is the mass of baryons.

If all these arguments are taken seriously, then (1) there must be *both* baryonic DM (to provide the DM in the Solar neighbourhood) and non-baryonic DM [so that $\Omega_0 = 1$ without violating Eq. (1.41)]; (2) the DM in galaxies may be baryonic but the DM in clusters like Coma must be non-baryonic (unless h is as small as 0.5, in which case the upper limit to M_B/L implied by nucleosynthesis may be barely consistent with the M/L's of rich clusters); (3) the ratio of DM to luminous mass must be larger outside galaxies, groups, and clusters than inside, since the mass-to-light ratios of these systems are not sufficient to close the Universe.

The candidates for non-baryonic DM in the context of particle physics are discussed by H. Harari in his lectures.

2. Theory of Stellar Dynamics

In most cases the determination of masses of galaxies and systems of galaxies is based on the dynamical theory of stellar systems. Here by "stellar systems" we mean systems composed of self-gravitating point masses, which may be either stars or galaxies. For the following lectures, we will restrict ourselves to spherical systems, although many of the objects we are interested in are not spherical. The main reason for using the spherical approximation is simplicity. Also the potential distribution is much more round than the density distribution so that even for flattened mass distributions a spherical potential is often not a bad approximation.

2.1. *Collisionless Boltzmann Equation*

We will first consider the equation of continuity for a fluid, and draw the analogy to the dynamics of discrete point masses in the next paragraph. The rate of change of mass within a volume V with surface S is

$$\frac{dM}{dt} = \int_V \frac{\partial \rho}{\partial t} d^3\vec{r} = -\int_S \rho\vec{v} \cdot d^2\vec{S}, \qquad (2.1)$$

where $\rho(\vec{r}, t)$ is the density distribution and $\vec{v}(\vec{r}, t)$ is the velocity field. Using Gauss's theorem,

$$\int_S \rho\vec{v} \cdot d^2\vec{S} = \int \nabla \cdot (\rho\vec{v}) d^3\vec{r}, \qquad (2.2)$$

so that Eq. (2.1) can be written,

$$\frac{dM}{dt} = \int_V \frac{\partial \rho}{\partial t} d^3\vec{r} = -\int_V \nabla \cdot (\rho\vec{v}) d^3\vec{r}. \qquad (2.3)$$

Since the equation must hold for an arbitrary volume in the absence of source and sink terms, we have

$$\frac{\partial \rho}{\partial t} + \nabla \cdot (\rho\vec{v}) = 0, \qquad (2.4)$$

which is the usual equation of continuity for fluids.

The dynamics of a system of point masses can be conveniently described by employing the phase space density distribution $f(\vec{r}, \vec{v}; t)$, where $f(\vec{r}, \vec{v}; t)d^3\vec{r}d^3\vec{v}$ is either the number, luminosity or mass contained in a phase space volume $d^3\vec{r}d^3\vec{v}$. Then one can derive a similar equation to the equation of continuity for the phase space density distribution using the same particle conservation argument in six-dimensional phase space rather than in normal three-dimensional space:

$$\frac{\partial f}{\partial t} + \sum_{i=1}^{3} \left[\frac{\partial}{\partial x_i}(f\dot{x}_i) + \frac{\partial}{\partial v_i}(f\dot{v}_i) \right] = 0. \qquad (2.5)$$

Now notice that $\dot{x}_i = v_i$, and $\partial v_i/\partial x_i = 0$. Furthermore $\dot{v}_i = -\partial\Phi/\partial x_i$, where Φ is the gravitational potential, so that $\partial\dot{v}_i/\partial v_i = 0$. Thus we find

$$\frac{\partial f}{\partial t} + \sum_{i=1}^{3} \left(v_i \frac{\partial f}{\partial x_i} - \frac{\partial\Phi}{\partial x_i}\frac{\partial f}{\partial v_i} \right) = 0, \qquad (2.6)$$

which is called the Vlasov or collisionless Boltzmann equation. Now define a convective derivative operator

$$\frac{D}{Dt} \equiv \frac{\partial}{\partial t} + \sum_i \left(v_i \frac{\partial}{\partial x_i} - \frac{\partial\Phi}{\partial x_i}\frac{\partial}{\partial v_i} \right), \qquad (2.7)$$

which gives the rate of change of a quantity as seen by an observer moving with a given star. Thus the collisionless Boltzmann equation is simply

$$\frac{Df}{Dt} = 0. \qquad (2.8)$$

The collisionless Boltzmann equation states that the local phase space density as viewed by an observer moving with a given star is conserved. This is analogous to a phenomenon seen in a marathon race. In the beginning of the race the spatial density of runners is high but their speeds vary over a wide range. As the race progresses, runners whose speed is nearly the same stay together so that near the finish, the runners in any given location have a low spatial density but travel at nearly the same speed. Therefore, phase space density remains roughly constant throughout the race. (We have assumed that all the runners who started the race finish so there are no sinks, and that each runner travels at constant speed.)

2.2. *The Jeans Theorem*

If the stellar system is in a steady state, the partial derivative with respect to time in Eq. (2.6) vanishes. We define integrals of motion $I(\vec{x}, \vec{v})$ to be functions such that

$$\frac{d}{dt} I[\vec{x}(t), \vec{v}(t)] = 0, \tag{2.9}$$

along any trajectory $[\vec{x}(t), \vec{v}(t)]$. The integrals satisfy the relation,

$$\frac{dI}{dt} = \sum_{i=1}^{3} \left(\dot{x}_i \frac{\partial I}{\partial x_i} + \dot{v}_i \frac{\partial I}{\partial v_i} \right) = \sum_{i=1}^{3} \left(v_i \frac{\partial I}{\partial x_i} - \frac{\partial \Phi}{\partial x_i} \frac{\partial I}{\partial v_i} \right) = 0. \tag{2.10}$$

Therefore, the integrals satisfy the time-independent collisionless Boltzmann equation and thus the phase space density distribution is a function only of the integrals, i.e. $f = f(I_1, I_2, \ldots)$. This is known as the Jeans theorem. In general an arbitrary stellar system can have six different integrals. In spherical systems only four of these are important for stellar dynamics: the energy per unit mass, E, and three components of the angular momentum per unit mass, $\vec{L}$. If the system is spherically symmetric in all respects (that is, any variable depends only on the distance from the center), the distribution function depends only on E and the square of the angular momentum, L^2. Therefore, the general solution of the collisionless Boltzmann equation for a spherically symmetric stellar system is any function of the form $f(E, L^2)$.

The number density distribution in a spherical system can be computed from

$$\nu(r) = \int_{-\infty}^{\infty} dv_r \int_{0}^{\infty} 2\pi v_t dv_t f\left(\frac{1}{2} v_r^2 + \frac{1}{2} v_t^2 + \Phi(r), r^2 v_t^2 \right), \tag{2.11}$$

where v_t and v_r are the tangential and radial velocities, respectively. If f represents the luminosity density or mass density, we can replace $\nu(r)$ by $j(r)$ or $\rho(r)$, respectively. The gravitational potential Φ satisfies Poisson's equation

$$\frac{1}{r^2} \frac{d}{dr} \left[r^2 \frac{d\Phi(r)}{dr} \right] = 4\pi G \rho(r). \tag{2.12}$$

Ideal observations could give us, at best, the distribution of $I(\vec{R}, v_{\parallel})$, where $I(\vec{R}, v_{\parallel}) d^2\vec{R} dv_{\parallel}$ is the luminosity in the area $d^2\vec{R}$ and velocity interval $dv_{\parallel}$ at projected position $\vec{R}$ and line-of-sight velocity $v_{\parallel}$. Even this information — far more than we are able to glean from present-day observations — is insufficient to give the distribution function $f(E, L^2)$, since we do not know the potential Φ to use in Eq. (2.11). Therefore, the determination of f is always an intrinsically under-determined problem. One must make additional assumptions to get the distribution function and mass distribution [for example, $j(r) \propto \rho(r)$].

2.3. *Examples of Distribution Functions*

In general the distribution functions for real stellar systems are not well-known. However, one often can make reasonable models using known analytic distribution functions. We will give two very simple examples of such distribution functions.

2.3.1. *Plummer model*

This is a power-law model for the phase space distribution function

$$f(E, L^2) = \begin{cases} K|E|^{7/2} & \text{for } E < 0 \\ 0 & \text{for } E > 0 \end{cases}. \tag{2.13}$$

In this case, the density distribution becomes

$$\nu(r) = 4\pi K \int_0^{\sqrt{2|\Phi|}} \left(|\Phi| - \frac{1}{2}v^2\right)^{7/2} v^2 dv = 7\pi^2 \cdot 2^{-11/2} K|\Phi|^5. \tag{2.14}$$

Now assume that $\nu = \rho$, i.e. $\nu(r)$ satisfies Poisson's equation

$$\frac{1}{r^2}\frac{d}{dr}\left(r^2\frac{d|\Phi|}{dr}\right) = -4\pi G\nu = -4\pi G \cdot 7\pi^2 \cdot 2^{-11/2} K|\Phi|^5. \tag{2.15}$$

A solution is a potential of the form

$$\Phi = -\frac{\Phi_0}{\sqrt{1 + r^2/a^2}}, \tag{2.16}$$

where Φ_0 is the central potential and the characteristic length scale a satisfies the relation,

$$a^2 = \frac{3 \cdot 2^{7/2}}{7\pi^3 G K \Phi_0^4}. \tag{2.17}$$

We have normalized the potential so that it vanishes at infinity. The Plummer model is one of the simplest models of stellar systems. The density distribution extends to infinity, but the mass is finite. It is obvious from Eqs. (2.14) and (2.16) that the density falls off as r^{-5} at large radius. Generally speaking, such a rapid falloff of density at large r is not compatible with the observed brightness distribution of galaxies, which decays somewhat more slowly (r^{-3} to r^{-4}). The Plummer model has the same density distribution as a gaseous polytrope with polytropic index $n = 5$.

2.3.2. *Isothermal sphere*

Consider a distribution function that follows the Maxwell–Boltzmann law

$$f(E) = Fe^{-E/\sigma^2}. \tag{2.18}$$

Such a distribution is achieved in gases through relaxation by collisions. However, in galaxies, where the relaxation timescale due to two-body gravitational encounters is much longer than the Hubble time, there is no fundamental reason to reach such a state. Nonetheless, the cores of elliptical galaxies and bulges of spirals are often found to be fitted with this model very well.

The density distribution for this model is

$$\nu(r) = 4\pi F \int_0^\infty v^2 dv e^{-\Phi/\sigma^2} e^{-v^2/2\sigma^2} = F(2\pi\sigma^2)^{3/2} e^{-\Phi/\sigma^2}. \tag{2.19}$$

If we set $\nu = \rho$, and choose the potential at the center to be zero, the density distribution becomes

$$\rho(r) = \rho_0 e^{-\Phi/\sigma^2}. \tag{2.20}$$

By substituting Eq. (2.20) into Poisson's equation, we get

$$\frac{1}{r^2}\frac{d}{dr}\left(r^2\frac{d\Phi}{dr}\right) = 4\pi G\rho_0 e^{-\Phi/\sigma^2}. \tag{2.21}$$

If we introduce the dimensionless variables, $\psi = -\Phi/\sigma^2$, $r_0 = \sqrt{9\sigma^2/4\pi G\rho_0}$, $s = r/r_0$, we get the dimensionless equation for ψ

$$\frac{1}{s^2}\frac{d}{ds}s^2\frac{d\psi}{ds} = -9e^\psi, \tag{2.22}$$

which also should satisfy the boundary conditions,

$$\psi(0) = 0, \quad \psi'(0) = 0. \tag{2.23}$$

Figure 1 shows the density and surface density distribution for the solution of Eq. (2.22), which is called the isothermal sphere. The central surface density satisfies $\Sigma_0 \approx 2\rho_0 r_0$ (more precisely the constant is 2.018), and the volume density falls as r^{-2} at large r. The radius parameter r_0 is called the core radius and it roughly corresponds to the radius where the surface density becomes half of the central value.

However, the isothermal sphere is unrealistic in that the total mass is infinite since $\rho \propto r^{-2}$ at large radii. More realistic models can be obtained by decreasing the phase space density at high energy. Models of this type are known as King or Michie models, but we will not discuss them here since the truncation affects only the outer parts, while we are concerned mostly with the inner parts in fitting observations of elliptical galaxies.

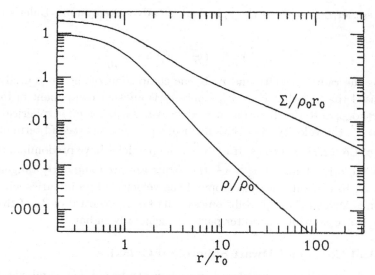

Fig. 1.   Density $\rho(r)$ and surface density $\Sigma(r)$ for an isothermal sphere, in units of the central density ρ_0 and core radius r_0.

2.4. *Moments of the Collisionless Boltzmann Equation*

By taking moments of the collisionless Boltzmann equation, we can get some useful equations relating observable quantities. The first step is just to integrate Eq. (2.6) over velocity space to get

$$\frac{\partial \nu}{\partial t} + \frac{\partial}{\partial x_i}(\nu \bar{v}_i) = 0, \tag{2.24}$$

where $\nu \bar{v}_i = \int f v_i d^3 \vec{v}$, and we have adopted the Einstein summation convention for simplicity of notation. The above equation is simply the equation of continuity (2.4). We can also multiply Eq. (2.6) by v_j and integrate over velocity space to obtain

$$\frac{\partial}{\partial t}(\nu \bar{v}_j) + \frac{\partial}{\partial x_i}(\nu \overline{v_i v_j}) + \frac{\partial \Phi}{\partial x_j}\nu = 0, \tag{2.25}$$

where $\nu \overline{v_i v_j} = \int f v_i v_j d^3 \vec{v}$. By introducing the stress tensor

$$\nu \sigma_{ij}^2 = \int f(v_i - \bar{v}_i)(v_j - \bar{v}_j) d^3 \vec{v} = \nu \overline{v_i v_j} - \nu \bar{v}_i \bar{v}_j, \tag{2.26}$$

and employing Eq. (2.24) we can rewrite Eq. (2.26) as

$$\frac{\partial \bar{v}_j}{\partial t} + \bar{v}_i \frac{\partial \bar{v}_j}{\partial x_i} = -\frac{\partial \Phi}{\partial x_j} - \frac{1}{\nu}\frac{\partial}{\partial x_i}(\nu \sigma_{ij}^2), \tag{2.27}$$

which is similar to Euler's equation for fluids. We can go further to obtain higher-order moment equations, but these are not generally useful for stellar dynamics.

If we restrict ourselves to the spherical, steady-state system, Euler's equation becomes,

$$\frac{d}{dr}(\nu \overline{v_r^2}) + \frac{2\nu}{r}(\overline{v_r^2} - \overline{v_\phi^2}) = -\nu \frac{d\Phi}{dr}, \tag{2.28}$$

where v_r is the radial velocity and v_ϕ is one component of the tangential velocity. We have used the fact that $\overline{v_\phi^2} = \overline{v_\theta^2}$, where v_θ is another component of tangential velocity orthogonal to the direction of ϕ, a necessary condition for spherical symmetry. If $\overline{v_\phi^2} = \overline{v_r^2}$, the velocity distribution is isotropic, and the second term of the left hand side of Eq. (2.28) vanishes. If $\overline{v_r^2} > \overline{v_\phi^2}$, the particles have predominantly radial orbits. On the other hand, if $\overline{v_\phi^2} > \overline{v_r^2}$, the orbits are predominantly tangential and near circular. In general, the anisotropy of the velocity ellipsoid varies with radius. If a system is formed through collisionless collapse, the central parts of the system tend to be isotropic while the outer parts are generally radial.

3. Elliptical Cores and Dwarf Spheroidal Galaxies

The theory developed in the previous lecture now can be applied to galaxies. We first attempt to determine mass-to-light ratios of cores of elliptical and dwarf spheroidal galaxies. To do this, we assume that the cores of these galaxies are (1) spherically symmetric and (2) have an isotropic velocity distribution. We further assume that (3) M/L is independent of radius. Finally we will assume that (4) the surface brightness distribution fits the isothermal distribution discussed in the previous lecture. By fitting the surface brightness profile to the isothermal sphere (Fig. 1) we can determine the core radius r_0 and the central surface brightness I_0. Then, Fig. 1 shows that the central emissivity is

$$j_0 \approx \frac{I_0}{2r_0}. \tag{3.1}$$

But by the definition of the core radius, the central density is

$$\rho_0 = \frac{9\sigma^2}{4\pi Gr_0^2}, \tag{3.2}$$

which can be used to determine M/L through

$$\frac{M}{L} = \frac{\rho_0}{j_0} = \frac{9\sigma^2}{2\pi GI_0r_0}. \tag{3.3}$$

One can generalize the above formula a little if one assumes that the cores of galaxies satisfy only the first three assumptions: spherical symmetry, isotropic velocity distribution and constant M/L. In this case, one can write the mass-to-light ratio as

$$\frac{M}{L} = \eta \cdot \frac{9\sigma_p^2}{2\pi GI_0r_h}, \tag{3.4}$$

where σ_p^2 is the line-of-sight mean square velocity dispersion at the center, η is a constant that depends on the actual distribution function of the stellar system, and r_h is the radius at which the surface brightness drops to half of its central value. The constant η is found to be very insensitive to the details of the actual models. For example, $\eta = 0.971$ for a Plummer model. One finds empirically that $\eta = 1$ to within a few percent for most stellar systems satisfying the three assumptions that have density distributions resembling those of real galaxies (Richstone and Tremaine 1986). Equation (3.4) with $\eta = 1$ is known as King's core-fitting formula and is commonly used for the determination of the mass-to-light ratios in the cores of galaxies.

How accurate is the core-fitting formula likely to be? The most serious concerns are the assumptions of velocity isotropy and spherical symmetry. A spherical galaxy composed entirely of stars on circular orbits would have $\sigma_p = 0$ and hence $\eta \to \infty$. However, such models are unrealistic; in the more plausible case of radial anisotropy $(\overline{v_r^2} > \overline{v_\phi^2})$, the constant η can be as small as 0.65 (Merritt 1987b). For non-spherical models η can be as low as 0.4 if the galaxy is viewed along its long axis (Merritt 1987b). The sensitivity to shape and velocity anisotropy is reduced if the dispersion is averaged over the central core rather than being measured precisely at the center.

Despite these concerns, in the absence of information on the orientation and velocity anisotropy of a given galaxy, the core-fitting formula (3.4) with $\eta = 1$ offers the best available estimate of the central M/L.

This technique has been applied to the cores of elliptical galaxies and bulges of spiral galaxies by Kormendy (1987a). No systematic differences are found between bulges and elliptical cores. The median value for the elliptical cores is found to be $M/L \approx 12\,h$, which is similar to the mass-to-light ratio expected from the Solar neighbourhood if gas and young stars are removed. This suggests that the DM in the Solar neighbourhood is also present in ellipticals and probably is a normal component of the stellar population.

The same technique has been applied to dwarf spheroidal galaxies by Aaronson and Olszewski (1987). There are about a half dozen low mass dwarf spheroidals that are satellites of our own Galaxy. Such galaxies are good places to look for DM since the density of luminous matter is very low. However, there are some difficulties in obtaining reliable velocity dispersion data for dwarf spheroidals since the velocity dispersion is low and the surface brightness is small. One has to measure accurate velocities of a number of individual stars to get a reliable velocity dispersion.

The velocity dispersion may be misleading if the measured stars are members of binary systems. To avoid contamination due to binaries, Aaronson and Olszewski measured the velocities of individual stars more than once to see if there is any velocity variation over the observation interval (typically one year). Any stars showing velocity variation are excluded in obtaining the velocity dispersion. The resulting M/L's for the dwarf spheroidals are listed in Table 2, as determined by Eq. (3.4) with $\eta = 1$. Notice that the stars used in estimating σ_p are spread out over the galaxy rather than being concentrated at the center; this implies that our estimates of M/L will be systematically low, by up to a factor of two or so.

Table 2. Mass-to-light ratios of dwarf spheroidal galaxies.

Name	N	$\sigma_p(\mathrm{km\,s^{-1}})$	$I_0(\mathrm{L_\odot pc^{-2}})$	$r_h(\mathrm{kpc})$	M/L
Fornax	5	6.4 ± 2	16.2	0.50	1.7
Sculptor	3	5.8 ± 2.4	9.7	0.17	6.8
Carina	6	5.6 ± 1.6	4.0	0.22	12
Ursa Minor	7	11 ± 3	1.2	0.15	220
Draco	9	9 ± 2	2.5	0.15	71

Notes: N denotes the number of stars used to determine the velocity dispersion. Parameters from Kormendy (1987b).

The M/L's of some dwarf spheroidals are normal (similar to the Solar neighbourhood or central parts of elliptical galaxies), but two, Draco and Ursa Minor, show very large M/L. If these large M/L's are correct, they show that some dwarf spheroidals are composed mostly of dark matter.

It has often been asked whether the exceptionally large M/L for Draco and Ursa Minor could be due to binary stars. Thus, one needs to know how much velocity dispersion can be attributed to the orbital motion of binaries, given the selection criterion against velocity variation employed by Aaronson and Olszewski. To address this question quantitatively, we have made a simple simulation. We have calculated the velocity dispersion due to orbital motions of binaries assuming that

(1) all stars are in binary systems;
(2) the primary component has mass $M_1 = 0.8\,\mathrm{M_\odot}$ and the cumulative mass distribution of the secondaries is $\propto M_2^{0.4}$ with $M_2 < M_1$, similar to the distribution for binaries in the Solar neighbourhood;
(3) the binary periods P are uniformly distributed in $\log(P)$ over a interval ± 0.25, centered on some value $\log(P_0)$; and
(4) the eccentricity distribution is uniform in e^2, as expected in statistical equilibrium.

Stars showing velocity variation $\geq 4\,\mathrm{km\,s^{-1}}$ between two epochs separated by 1 year are excluded in computing the line-of-sight velocity dispersion since the observers exclude them as well. Figure 2 shows the line-of-sight velocity dispersion σ_p as a function of period P_0. Also shown is the ratio f_v of the number of stars showing velocity variation to the number showing no variation. This graph shows that there is a good correlation between the velocity dispersion due to binary systems and the fraction of stars showing velocity variation. Therefore, once the fraction of stars showing velocity variation among the sample is known, one can estimate the contribution to the velocity dispersion from binary orbital motions.

This result can be applied to the two observed dwarf spheroidals with high M/L's. The Draco system has 11 stars observed with radial velocities. Two of them show velocity variation, so $f_v \approx 0.2$. From Fig. 2, the velocity dispersion contribution due to binaries is then at most about $2\,\mathrm{km\,s^{-1}}$. The contribution of

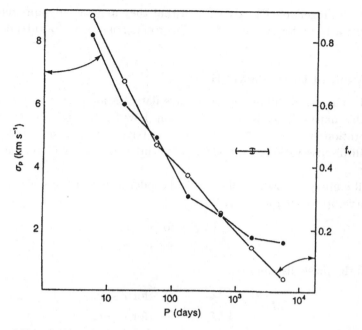

Fig. 2. Plot of line-of-sight velocity dispersion σ_p due to binaries as a function of binary period, as well as the ratio f_v of the number of stars showing velocity variation exceeding $4\,\mathrm{km\,s^{-1}}$ over a 1 year interval to the number showing variation $< 4\,\mathrm{km\,s^{-1}}$. We have assumed that all stars are in binaries with $\log(P)$ uniformly distributed over the interval ± 0.25 around the plotted value. The error bars indicate the width of the assumed distribution in $\log(P)$ and the statistical uncertainty in the results. The curves are not completely smooth because of resonance effects between the mean binary period and the observation interval.

binaries to the observed dispersion of $9\,\mathrm{km\,s^{-1}}$ is therefore less than about 5% in quadrature. The same analysis for Ursa Minor, which has a dispersion of $11\,\mathrm{km\,s^{-1}}$ from 7 stars (3 out of 10 measured stars show velocity variation, so $f_v \approx 0.43$), shows that the dispersion due to binaries is about $4\,\mathrm{km\,s^{-1}}$ so that the observed velocity dispersion is in error by at most 13%. Similar results are obtained for other period distributions. Therefore it may safely be concluded that the effects of binaries are negligible in Draco and Ursa Minor.

Are there other effects that might contribute to erroneous M/L's? The radial velocity data could be contaminated by motions of the stellar atmospheres. However, K-giants, which are the main sources of velocity determinations, have very small atmospheric motions. The core-fitting formula (3.4) is already biased toward low M/L's because it is based on the central velocity dispersion while the measured stars often lie outside the core where the dispersion is lower. Statistical errors are also biased toward smaller M/L since the χ^2 distribution is asymmetric.

Perhaps the most interesting alternative to the existence of large quantities of dark matter is the possibility that Draco and Ursa Minor are unbound systems. Their extremely low surface brightness, and their location in the plane of the orbit of the Magellanic Clouds, suggest that they may be only apparent density enhancements arising, for example, from the crossing of streams of tidal debris from the Clouds.

4. The Extent of the Galactic Halo

Most spiral galaxies, including our own, have flat rotation curves as far as one can measure. This naturally leads to the question, how far beyond the last measured point do rotation curves stay flat, that is, how far do dark galactic halos extend? Here we will discuss various methods to determine the extent of the dark halo of our Galaxy.

We shall employ a very simple spherical model in which the Galaxy has a flat rotation curve up to the radius r_*; thus

$$V_{\text{rot}}^2 = \begin{cases} V_0^2 & \text{for } r < r_*, \\ \propto \frac{1}{r} & \text{for } r > r_* \end{cases}, \tag{4.1}$$

which gives the mass distribution

$$M(r) = \begin{cases} \frac{V_0^2 r}{G} & \text{for } r < r_* \\ M_* = \frac{V_0^2 r_*}{G} & \text{for } r > r_* \end{cases}. \tag{4.2}$$

Now our aim is to determine the value of r_*. Rotation curve measurements show constant rotation velocity up to $r \approx 2R_0$, where R_0 is the Galactocentric distance of the sun; thus $r_* \gtrsim 2R_0 \approx 17\,\text{kpc}$.

4.1. Local Escape Speed

The velocity distribution of stars in the Solar neighbourhood with respect to the rest frame of our Galaxy shows a cut-off near $v_{\text{max}} \approx 500\,\text{km\,s}^{-1}$ (Carney and Latham 1987). This is interpreted as a lower limit to the escape speed of stars in the Solar neighbourhood. This means that the following equation holds,

$$\frac{1}{2}v_{\text{max}}^2 + \Phi(R_0) < 0. \tag{4.3}$$

The gravitational potential satisfies the equation,

$$\frac{d\Phi}{dr} = \frac{GM(r)}{r^2}, \tag{4.4}$$

which in the model of Eqs. (4.1) and (4.2) can be integrated to yield

$$\Phi = \begin{cases} V_0^2 \left[\ln \frac{r}{r_*} - 1\right] & \text{for } r < r_* \\ -\frac{V_0^2 r_*}{r} & \text{for } r > r_* \end{cases}, \tag{4.5}$$

where we have normalized the potential to vanish at infinity. By plugging $V_0 = 220\,\text{km\,s}^{-1}$ and $v_{\text{max}} = 500\,\text{km\,s}^{-1}$ into Eq. (4.3), we get $r_* \gtrsim 4.9R_0$. If we further use $R_0 \approx 8.5\,\text{kpc}$, the extent of the halo of our galaxy is roughly $r_* \gtrsim 41\,\text{kpc}$.

This corresponds to a mass $M_* \gtrsim 4.6 \times 10^{11} M_\odot$. Since the total luminosity of our galaxy at the V band is $L_V \approx 1.4 \times 10^{10} L_\odot$, the mass-to-light ratio becomes

$$\frac{M}{L} \gtrsim 33, \tag{4.6}$$

at least a factor of six larger than the M/L in the Solar neighbourhood.

4.2. Magellanic Stream

The Large Magellanic Cloud (LMC: Galactocentric distance $d = 52\,\text{kpc}$ and $M \approx 2 \times 10^{10}\,M_\odot$) and Small Magellanic Cloud (SMC: $d = 63\,\text{kpc}$ and $M \approx 2 \times 10^9\,M_\odot$) are the two nearest satellite galaxies to us. They contain $\approx 5 \times 10^9\,M_\odot$ of neutral hydrogen gas. In addition, there is a stream or trail of HI extending away from the clouds. This HI trail follows a great circle and contains almost $10^9\,M_\odot$ of neutral hydrogen. It is presumably composed of gas from the Clouds that has been stripped off by the tidal field of the Galaxy.

Most of this gas is probably on free Kepler orbits, not too dissimilar from that of the Clouds. At the tip of the stream, gas is falling towards us at high speed ($v \approx -220\,\text{km}\,\text{s}^{-1}$ in the Galactic rest frame). If the Galactic potential were that of a point mass, this large velocity would suggest that the material at the tip has fallen deep into the potential, to a Galactocentric distance $\lesssim 15\,\text{kpc}$. However, parallax effects due to the offset of the Sun from the Galactic center should then spoil the great circle shape of the stream. This argument suggests that there is a massive halo in our Galaxy, since then the required infall velocity can be achieved at a larger radius.

Detailed dynamical models have been constructed by Murai and Fujimoto (1980) and Lin and Lynden-Bell (1982). If a flat rotation curve is assumed for our Galaxy, the best fitting circular velocity is $244 \pm 20\,\text{km}\,\text{s}^{-1}$, and the rotation curve must remain flat to at least 70 kpc. This suggests that there is a very large amount of DM in our Galaxy, much more than the lower limit obtained above using the local escape velocity. The mass-to-light ratio of the Galaxy would exceed 56 in Solar units if such an extended halo were present.

4.3. Local Group Timing

The nearest giant spiral galaxy M31 is about 730 kpc away. Together with the associated satellite galaxies, we and M31 compose a relatively isolated system of galaxies, known as the Local Group. The relative radial velocity of the center of mass of us and M31 is $-119\,\text{km}\,\text{s}^{-1}$, which has the opposite sign to the usual Hubble flow. It is possible that the two galaxies are approaching each other by chance. However, a more natural explanation is that the two galaxies were once moving apart due to the Hubble expansion, but that the expansion was slowed and reversed by their mutual gravitational attraction. If we assume M31 and Galaxy to be point masses and ignore the presumably small masses of other Local Group members, the relative orbit of the two galaxies can be written parametrically as

$$r = a(1 - e\cos\eta), \quad t = \sqrt{\frac{a^3}{GM}}(\eta - e\sin\eta) + C, \tag{4.7}$$

where C is a constant, a is the semi-major axis, e is the eccentricity, η is the eccentric anomaly, and M is the total mass of the Local Group [compare Eq. (1.34)]. Since the two galaxies were presumed to be at the same place at the beginning of the expansion, $r = 0$ at $t = 0$ so that we have to choose $e = 1$ and $C = 0$, corresponding to a radial orbit. The relative radial velocity may be written

$$\frac{dr}{dt} = \sqrt{\frac{GM}{a}}\frac{\sin\eta}{1 - \cos\eta} = \frac{r}{t}\frac{\sin\eta(\eta - \sin\eta)}{(1 - \cos\eta)^2}. \tag{4.8}$$

By plugging in $dr/dt = -119\,\mathrm{km\,s}^{-1}$ and $t = 10$–$20\,\mathrm{Gyr}$ ($1\,\mathrm{Gyr} = 10^9$ years), we get $\eta = 4.11$ (for $t = 10\,\mathrm{Gyr}$) to 4.46 ($t = 20\,\mathrm{Gyr}$) radians. Then the mass of the Local Group is found to be $M = 5.5 \times 10^{12}\,\mathrm{M}_\odot$ (for $t = 10\,\mathrm{Gyr}$) or $3.2 \times 10^{12}\,\mathrm{M}_\odot$ (for $t = 20\,\mathrm{Gyr}$). This gives the mass-to-light ratio of the Local Group

$$\frac{M}{L} = 76 - 130, \tag{4.9}$$

which is again very large. We expect that this roughly represents the M/L of our Galaxy if there is no big variation of M/L between M31 and the Galaxy.

4.4. *Kinematics of Satellite Galaxies*

In principle, the kinematics of satellite galaxies and globular clusters can provide a useful tool to determine the mass of the Galaxy. However, this determination is intrinsically difficult since there are only a few objects and only the line-of-sight velocities are known. In particular, the satellite galaxies are at distances large compared to the distance to the Galactic center from the Sun, so we see mainly the radial component of their velocity. Thus we must make some statistical assumption about the ratio of tangential to radial velocity.

If we assume r_* is large compared to the distances of the sample objects, the gravitational potential (4.5) has the form

$$\Phi(r) = V_0^2 \ln r + \text{constant}. \tag{4.10}$$

The moment of inertia per unit mass is

$$I = r^2, \tag{4.11}$$

and its first and second derivatives are

$$\dot{I} = 2\vec{r}\cdot\dot{\vec{r}}; \quad \ddot{I} = 2\vec{r}\cdot\ddot{\vec{r}} + 2v^2. \tag{4.12}$$

By taking an average over several galaxies, we get

$$\langle v^2 \rangle = \left\langle r\frac{d\Phi}{dr} \right\rangle = V_0^2. \tag{4.13}$$

Table 3. Globular clusters and satellite galaxies used for Galactic mass determination.

Name	$d_\odot$(kpc)	$v_\odot$(km s^{-1})[†]	v_G(km s^{-1})[§]
LMC + SMC	52	245 ± 5	62
Draco	75	-289 ± 1	-95
Ursa Minor	63	-249 ± 1	-88
Sculptor	79	107 ± 2	75
Fornax	138	55 ± 5	-34
Carina	91	230 ± 1	14
AM1	116	116 ± 15	-42
NGC 2419	90	-20 ± 5	-26
Pal 3	91	89 ± 9	-59
Pal 4	105	75 ± 5	54

Notes The list contains all known satellites with $d_\odot > 50$ kpc and published velocity error < 20 km s^{-1}. Data sources given by Little and Tremaine (1987).
†: radial velocity with respect to the Sun
§: radial velocity with respect to the Galactic center

If the velocity distribution is isotropic, $\langle v_r^2 \rangle = \frac{1}{3}\langle v^2 \rangle = \frac{1}{3}V_0^2$. For $V_0 = 220$ km s^{-1}, this gives $\langle v_r^2 \rangle^{1/2} = 127$ km s^{-1}, while the actual data exhibit $\langle v_r^2 \rangle^{1/2} = 60$ km s^{-1} (Table 3). This means that the existing data are consistent with a flat rotation curve extending to very large distances only if the velocity distribution is primarily tangential, a conclusion first reached by Lynden-Bell et al. (1983). Dissipationless collapse leads to galaxies with a velocity distribution that varies from radial to isotropic, and it is difficult to construct galaxies with predominantly tangential velocities. Thus we are led to question the existence of an extended massive halo for the Galaxy.

There have recently been substantial improvements in the accuracy of velocity data for satellite galaxies. Furthermore, it is unlikely that the quality or quantity of radial velocity data will be improved drastically in the near future. Therefore it is perhaps appropriate to elaborate on the arguments above and to make a careful statistical model of the currently available kinematic data for satellites (Little and Tremaine 1987).

Let us introduce a fictitious mass for an observed satellite i

$$\mu_i \equiv \frac{v_{r,i}^2 r_i}{G},$$
(4.14)

where r_i and $v_{r,i}$ are the distance from the Galactic center and the radial velocity with respect to the rest frame of the Galaxy, respectively. For a Galactic mass M, let us denote $P(\mu_i|M)d\mu$ as the probability of finding μ_i in the range between μ_i and $\mu_i + d\mu$, which can be computed from

$$P(\mu|M) = \frac{\int d^3\vec{r}\, d^3\vec{v}\, f(\vec{r}, \vec{v})\delta(\mu - v_r^2 r/G)}{\int d^3\vec{r}\, d^3\vec{v}\, f(\vec{r}, \vec{v})},$$
(4.15)

where $f(\vec{r}, \vec{v})$ is the usual phase space density distribution. Two different models for the Galactic potential, a point mass and an infinite halo, and two different models for the velocity distribution, radial and isotropic, are considered. For the point mass potential,

$$P(\mu|M) = \begin{cases} \frac{2}{5\pi} \frac{\max[0,(2M-\mu)]^{5/2}}{\mu^{1/2}M^3} & \text{for isotopic orbits} \\ \frac{1}{\pi} \frac{\max[0,(2M-\mu)]^{1/2}}{\mu^{1/2}M} & \text{for radial orbits} \end{cases}. \qquad (4.16)$$

Now what we want to know is the corresponding probability distribution for M given μ, that is, $P(M|\mu)$. The relation between $P(\mu|M)$ and $P(M|\mu)$ is given by Bayes's theorem

$$P(M|\mu) = \frac{P(\mu|M)P(M)}{\int P(\mu|M')P(M')dM'}, \qquad (4.17)$$

where $P(M)dM$ is the *a priori* probability for a galaxy to have mass between M and $M + dM$. If several objects are available the probability becomes

$$P(M|\mu_1,\ldots,\mu_N) = P(M)\frac{\prod_{i=1}^{N} P(\mu_i|M)}{\int \prod_{i=1}^{N} P(\mu_i|M')P(M')dM'}. \qquad (4.18)$$

We now have to specify $P(M)$ to proceed with our analysis. One of the most reasonable choices is that the *a priori* distribution is uniform in $\log M$ so that

$$P(M)dM \propto d\log M \quad \text{or} \quad P(M) \propto \frac{1}{M}. \qquad (4.19)$$

This choice is obviously somewhat arbitrary. However, the results become less and less sensitive to the choice of the functional form of $P(M)$ as the number of data points increases, and for the numbers we are dealing with ($N \approx 10$) the choice of $P(M)$ has no strong influence on the results.

The data set has been gathered from various sources. All globular clusters and satellite galaxies at distances $d > 50\,\text{kpc}$ with velocity errors less than $20\,\text{km\,s}^{-1}$ are used. The data are listed in Table 3. The LMC and SMC are treated as one data point at their center of mass since the motion of each Cloud is greatly influenced by the other Cloud.

Using Eqs. (4.16) and (4.18) we find that for isotropic orbits there is a 90% probability that the Galactic mass M lies in the range $[1.4, 5.2] \times 10^{11}\,\text{M}_\odot$ with the median value being $2.4 \times 10^{11}\,\text{M}_\odot$. For radial orbits the mass is even smaller. If we define r_* so that the mass obtained above corresponds to $V_0^2 r_*/G$ with $V_0 = 220\,\text{km\,s}^{-1}$, we get $r_* \leq 46\,\text{kpc}$ at the 95% confidence level for isotropic orbits. This result is self-consistent in that the upper limit to r_* is smaller than the lower limit of $50\,\text{kpc}$ for the data so that the point mass approximation is valid over most of the orbit of a typical satellite.

A similar analysis has been made assuming an infinite halo potential. In this case, the unknown is the circular speed V_0 rather than the total mass of the Galaxy. Here we find the range of circular velocity at the 90% confidence level

Table 4. Values of r_* from various methods.

Method	r_*(kpc)
Direct measurement of rotation curve	$\gtrsim 20$
Local escape velocity	$\gtrsim 40$
Magellanic Stream	$\gtrsim 80$
Local Group timing	≈ 100
Kinematics of satellites	≈ 50

to be $[77, 165]\,\mathrm{km\,s^{-1}}$, with the median value being $107\,\mathrm{km\,s^{-1}}$. This is clearly too small compared to the local speed of $220\,\mathrm{km\,s^{-1}}$, and once again implies that the Galaxy's massive halo has only a limited extent.

This analysis strongly suggests that the Galactic halo does not extend much beyond the outermost visible components of the Galaxy at $r \approx 30\,\mathrm{kpc}$. However, it should be noted that the value for r_* obtained here is considerably lower than the values obtained from the Magellanic Stream and Local Group timing. If some unknown formation mechanism has placed the satellites on orbits with predominantly tangential velocities, then the satellite kinematics could be consistent with these other arguments.

4.5. *Summary*

We summarize the various results for the determination of the Galactic mass in Table 4.

All of these estimates agree that the Galactic halo is much more extended than the bulk of the visible stars and gas. Hence most of the mass of the Galaxy is dark. However, there are some contradictions between various determinations of the Galactic mass, particularly between the satellite kinematics on the one hand and the Magellanic Stream and Local Group timing on the other. The source of this discrepancy remains mysterious. Perhaps (1) the gas in the Magellanic Stream is not on free Kepler orbits; (2) the Galaxy and M31 are embedded in a low-density mass concentration so that most of the mass in the Local Group is not associated with either galaxy; or (3) the satellite galaxies and distant globular clusters are on nearly circular orbits.

Thus, it is still unclear whether the halo of the Galaxy — and by analogy, the halos of other spiral galaxies — extends only to 50 kpc or so, about twice the optical extent, or out to much larger distances, up to several hundred kiloparsecs.

5. Binary Galaxies

Carefully selected samples of binary galaxies provide another opportunity to determine the mass of individual galaxies. However, constructing a well-defined sample is not an easy task. As an example, we discuss the sample used by Turner (1976).

From the Zwicky catalogue of galaxies, he restricted his sample to galaxies in the northern hemisphere (i.e. declination $\delta > 0$) in order to ensure completeness. To avoid heavy Galactic extinction, he also selected only galaxies at Galactic latitude $|b| > 40°$. The flux limit for his sample was taken to be 15th magnitude, again to help ensure completeness. The selection criteria for binary galaxies were: (1) the separation of the binary (θ_{12}) does not exceed 8 arcminutes; (2) the next nearest galaxy should lie beyond $5\theta_{12}$. The second criterion was used to avoid possible contamination from groups and clusters. The final number of binary galaxies was 156 pairs out of some 30,000 galaxies in Zwicky catalogue.

Despite these stringent selection criteria, later close visual examination by White et al. (1983) showed that many of Turner's pairs are members of larger clusters or groups. There are two reasons why Turner's selection criteria prove to be insufficiently stringent. First, groups or clusters can easily have one or two bright members that are included in Turner's sample, while the other group members fall below the magnitude limit. Second, if two galaxies happen to be nearby in the plane of the sky, there may well be no other galaxy within five times their separation even in rich clusters. White et al. find that only 76 pairs out of Turner's 156 survived the additional culling of all pairs in visible clusters or groups. Clearly, it is very difficult to construct a large, well-defined sample of isolated binaries.

If we ignore this concern, we can carry out some analysis using the existing sample. We shall work with the moment Eq. (2.28) for spherical systems. If we assume a flat rotation curve so that the potential $\Phi(r) = V_c^2 \ln r + \text{const}$, and the degree of anisotropy $\beta = 1 - \overline{v_\phi^2}/\overline{v_r^2}$ is a constant independent of r, then Eq. (2.28) reduces to

$$\frac{d}{dr}(\nu\overline{v_r^2}) + \frac{2\overline{v_r^2}\nu}{r}\beta = -\frac{\nu V_c^2}{r}. \tag{5.1}$$

The statistical analysis of large catalogs of galaxies shows that the galaxies have a two-point correlation function $\xi \propto r^{-\gamma}$, with $\gamma \approx 1.8$. This implies that the distribution of binary galaxies follows $\nu \propto r^{-\gamma}$. A solution to Eq. (5.1) is then

$$\overline{v_r^2} = \text{const} = V_c^2/(\gamma - 2\beta).$$

If the system has an isotropic velocity dispersion tensor ($\beta = 0$), we get $V_c = 170 \, \text{km s}^{-1}$ after plugging in the observed rms velocity difference $(\overline{v_\parallel^2})^{1/2} = 127 \, \text{km s}^{-1}$ from Turner's sample ($\overline{v_\parallel^2} = \overline{v_r^2}$ since the velocities are isotropic). This is more or less consistent with other determinations. However, if the orbits are radial ($\beta = 1$), no solution with $\overline{v_r^2}$ and V_c^2 positive is possible. In general, we can get any value of $V_c < 170 \, \text{km s}^{-1}$ by adjusting the unknown parameter β between 0 and $0.5\gamma = 0.9$; thus the results are entirely dependent on the unknown anisotropy of the orbits.

Do we learn anything from binary galaxies? Because of the problems we have mentioned above, the answer is "not much". Perhaps the most interesting result so far is that the velocity difference between the galaxies does not correlate either with the projected distance between binaries or the luminosity. This suggests that mass is not related to the luminosity of the host galaxy.

6. Masses of Groups and Clusters of Galaxies

6.1. *Groups of Galaxies*

The virial theorem discussed early in these lectures can be used to determine the masses of groups or clusters of galaxies in much the same way Zwicky did. The virial theorem [Eq. (1.8)] states

$$\sum_{i=1}^{N} m_i v_i^2 - \frac{G}{2} \sum_{\substack{i,j=1 \\ i \neq j}}^{N} \frac{m_i m_j}{|\vec{r}_i - \vec{r}_j|} \approx 0, \tag{6.1}$$

where we have used $\approx$ rather than $=$ because the equation is only precisely true in a time averaged sense. We now take an average of Eq. (6.1) over angle, denoted by $\langle \rangle_\Omega$. We have

$$v_i^2 = 3 \langle v_{\|,i} \rangle_\Omega, \tag{6.2}$$

where $v_\|$ denotes the line-of-sight velocity. The projected distances $\vec{R}_i$ are likewise related to the three-dimensional distances $\vec{r}_i$ through

$$\left\langle \frac{1}{|\vec{R}_i - \vec{R}_j|} \right\rangle_\Omega = \frac{1}{|\vec{r}_i - \vec{r}_j|} \left\langle \left| \frac{1}{\sin \theta_{ij}} \right| \right\rangle_\Omega = \frac{\pi}{2} \frac{1}{|\vec{r}_i - \vec{r}_j|}, \tag{6.3}$$

where we have assumed that the direction of a vector $\vec{r}$ is random. The averages in Eqs. (6.2) and (6.3) are not meaningful if one looks at only one or two galaxies. However, these relations can be used to estimate average three-dimensional velocities and separations when summed over many galaxies.

Thus we may write Eq. (6.1) as

$$3 \sum_{i=1}^{N} m_i v_{\|,i}^2 - \frac{G}{\pi} \sum_{\substack{i,j=1 \\ i \neq j}}^{N} \frac{m_i m_j}{|\vec{R}_i - \vec{R}_j|} \approx 0. \tag{6.4}$$

We now have several different options for determining the mass of the system: (1) we may assume that $m_i \approx (M/L)L_i$ (i.e. the mass-to-light ratio is constant for all galaxies) or (2) we may assume that $m_i = M/N$, where N is the total number of galaxies and M is the total cluster mass (i.e. the number distribution of galaxies in phase space is proportional to the distribution of mass). If we take the second assumption, the total mass-to-light ratio may be written

$$\frac{M}{L} = \frac{3\pi N}{G} \frac{\sum_i v_{\|,i}^2}{\sum_k L_k \sum_{i \neq j} 1/|\vec{R}_i - \vec{R}_j|}. \tag{6.5}$$

Several alternative methods to the virial theorem for estimating masses of groups of galaxies have been proposed by Heisler et al. (1985). For example, there is the "median mass" estimator

$$M_{\rm Me} = \frac{f_{\rm Me}}{G} \text{med}_{i,j}\left[(v_{\|,i} - v_{\|,j})^2 |\vec{R}_i - \vec{R}_j|\right], \qquad (6.6)$$

where $\text{med}_{i,j}$ denotes the median over all pairs of galaxies (i,j). The proportionality constant $f_{\rm Me}$ is determined to be approximately 6.5 from numerical experiments.

The median mass estimator is found to give similar masses to the virial theorem. No one method appears to be superior to another, and in a given case it appears that all known estimators tend to err in the same direction.

These mass estimators are applied to the catalog of groups of galaxies compiled by Huchra and Geller (1982). The median mass-to-light ratios are found to be approximately $400\,h$ in Solar units with errors being about ± 0.4 in the logarithm between the median and the quartiles (that is, a factor of 3). Thus there is clear evidence for large quantities of DM in groups.

6.2. *Rich Clusters: Coma Cluster*

We can make more elaborate models for rich clusters than for groups, since there are many more galaxies available. For the sake of definiteness, we will concentrate on the Coma cluster, which provides one of the best examples of the evidence for DM. The virial theorem is not useful in this case because it is plagued by issues of cluster membership in the outer parts. The core-fitting method discussed in lecture 3 is not particularly useful either, in part because there is no well-defined core in the Coma cluster.

The moment Eq. (2.28) can be written

$$\frac{d}{dr}(\nu \overline{v_r^2}) + \frac{2\nu \overline{v_r^2}}{r}\beta(r) = -\nu \frac{d\Phi}{dr} = -\nu \frac{GM(r)}{r^2}, \qquad (6.7)$$

where the anisotropy parameter $\beta(r) = 1 - \overline{v_\phi^2}/\overline{v_r^2}$. If we assume spherical symmetry, we can determine the line-of-sight dispersion profile $v_\|^2(r)$ and the number density profile $\nu(r)$ from observations. However, Eq. (6.7) is still underdetermined, since it involves four functions, $\beta(r)$, $M(r)$, $\overline{v_r^2}(r)$, and $\nu(r)$, and we have only three constraints (the two observable quantities and the equation). Merritt (1987a) has made several assumptions to determine the mass of the cluster as follows:

(1) Number traces mass (i.e. $\nu \propto \rho$): With this assumption, the observed dispersion and number density profiles imply a nearly isotropic velocity distribution (i.e. $\beta \approx 0$). The determination of mass in this case is straightforward, and gives $M = 1.8 \times 10^{15}\,h^{-1}{\rm M}_\odot$.

(2) Minimum mass model: Such models can be obtained by concentrating the mass in the center as much as possible without violating observations. The density

profile is assumed to be

$$\rho(r) \propto \frac{1}{(1 + r^2/r_0^2)^{n/2}}. \tag{6.8}$$

Then Merritt attempts to make r_0 as small as possible. He finds rather implausible models in this case, that is, all the orbits are nearly circular outside the core. For these models the total mass never becomes lower than $0.7 \times 10^{15}\, h^{-1}\mathrm{M}_\odot$. The DM is mostly concentrated within two optical core radii.

(3) Maximum mass model: One can obtain a rather high mass model by distributing the DM more or less uniformly. The total mass and central density vary over a wide range. However, the mass within $1\, h^{-1}$ Mpc is within 25% of the value obtained for case (1).

(4) Radial dependence of anisotropy: One also can choose a functional form for the radial dependence of the velocity anisotropy,

$$\beta(r) = \frac{r^2}{r^2 + r_a^2}, \tag{6.9}$$

where r_a is a free parameter. This choice has some nice features, in particular, the orbits are more radial in the outer parts, as one expects from a collisionless collapse process. Furthermore, models of this kind have a known distribution function of the form

$$f = f\left(E + \frac{L^2}{2r_a^2}\right). \tag{6.10}$$

Models with the anisotropy radius r_a exceeding the optical core radius r_0 are found to be consistent with observations. This means that models having predominantly radial orbits outside the core radius, in which the DM is more uniformly distributed than the galaxies, are allowed.

In conclusion, the mass of the Coma cluster is about $2 \times 10^{15}\, h^{-1}\mathrm{M}_\odot$ if the dark matter is distributed like the galaxies; the mass cannot be less than 40% of this value but may be much more if the dark matter is more extended than the galaxies. For a total luminosity of $5 \times 10^{12}\, h^{-2}\mathrm{L}_\odot$, the corresponding mass-to-light ratio is about $400\, h$ in Solar units. The distribution of DM is not well-constrained, but it seems to be inescapable that most of the mass in the Coma cluster is dark.

There are several other methods to determine the masses of galaxies and galaxy systems, and we will close by mentioning two.

X-ray observations of elliptical galaxies have been used to determine the mass of these galaxies, assuming hydrostatic equilibrium for the X-ray emitting hot gas. The advantage of this technique is that the distribution function of the gas is known to be isotropic, so that the anisotropy parameter $\beta = 0$. However, existing spectral data, which are essential to determine the temperature profile, have no spatial resolution, and the results are sensitive to the assumed temperature profile. The sole exception, where adequate resolution is available, is the giant elliptical galaxy M87.

One problem in this case is that M87 is located in the center of the Virgo cluster so that its mass may not be typical of other galaxies. The analyses derive a very large mass-to-light ratio, at least 750 in Solar units (Fabricant and Gorenstein 1983). This technique will become more widespread when high-resolution spectral data are available from the AXAF satellite.

The perturbations to the Hubble flow induced by the nearby Virgo cluster of galaxies have been used to estimate the mass of the cluster. The mass-to-light ratio appears to be comparable to that of Coma and other rich clusters, but substantially less than the value required so that the Universe is closed.

7. Summary

The determinations of masses of galaxies and systems of galaxies on various scales show large variations in M/L. Generally M/L increases as the scale of the object increases. For example, $M/L \approx 5$ in the Solar neighbourhood (on a scale of a few

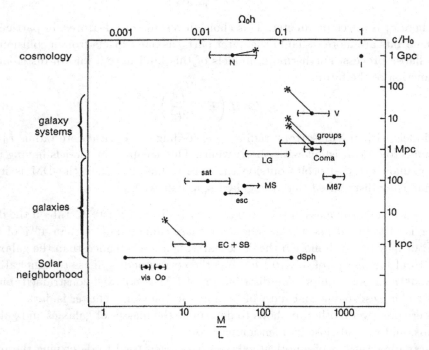

Fig. 3.   Mass-to-light ratios of galaxies and galaxy systems discussed in the text. The key to the labels is: "N", nucleosynthesis; "I", inflation; "V", Virgo flow; "groups", groups of galaxies; "Coma", Coma cluster of galaxies; "LG", Local Group timing; "M87", X-ray observations of M87 galaxy; "sat", satellites of our Galaxy; "MS", Magellanic Stream; "esc", escape speed from Solar neighbourhood; "EC+SB", elliptical cores and spiral bulges; "dSph", dwarf spheroidal galaxies; "vis", visible components of the Solar neighbourhood; "Oo", Oort limit. All values are plotted for $h = 1$, but a line with an asterisk at the end is used to show how results for M/L and scale change when $h = 0.5$. This is a revised version of a diagram due to Ostriker et al. (1974).

hundred parsecs) while $M/L \approx 400\,h$ in the Coma cluster which has a scale of several Mpc. One notable exception to this trend is that some dwarf spheroidal galaxies show large M/L on scales less than a kiloparsec. The mass density implied by these M/L's is never large enough for closure, which corresponds to an average $M/L \approx 1600\,h$ for the entire Universe.

Baryonic matter can account for the DM on scales up to the size of dark halos of galaxies, but non-baryonic mass is probably required to provide the dark mass in rich clusters unless the Hubble constant is as small as $50\,\mathrm{km\,s^{-1}\,Mpc^{-1}}$.

To summarize, we have prepared a schematic graph in Fig. 3, which shows the various determinations of M/L in different objects and on different scales.

Acknowledgements

Much of the material in these lectures is drawn from a book that Tremaine has written with James Binney (Binney and Tremaine 1987), and the treatment given here has been greatly influenced by our joint labours over the past several years. We are grateful to David Merritt for constructive suggestions and criticisms on a number of topics. Of course, the responsibility for errors is our own.

References

Aaronson, M. and Olszewski, E. (1987) In: *Dark Matter in the Universe, IAU Symposium No. 117*, p. 153, eds. Kormendy, J. and Knapp, G. R., Reidel, Dordrecht, The Netherlands.

Binney, J. J. and Tremaine, S. (1987) *Galactic Dynamics*, Princeton University Press, Princeton, NJ.

Carney, B. W. and Latham, D. W. (1987) In: *Dark Matter in the Universe, IAU Symposium No. 117*, p. 39, eds. Kormendy, J. and Knapp, G. R., Reidel, Dordrecht, The Netherlands.

Davis, M. and Huchra, J. (1982) *Astrophys. J.*, **254**, 425.

Faber, S. M. and Gallagher, J. S. (1979) *Ann. Rev. Astron. Astrophys.*, **17**, 135.

Fabricant, D. and Gorenstein, P. (1983) *Astrophys. J.*, **267**, 535.

Gunn, J. E. (1978) In: *Observational Cosmology, Eighth Advanced Course of the Swiss Society of Astronomy and Astrophysics*, p. 1, eds. Maeder, A., Martinet, L., and Tammann, G., Geneva Observatory, Geneva, Switzerland.

Heisler, J., Tremaine, S., and Bahcall, J. N. (1985) *Astrophys. J.*, **298**, 8.

Huchra, J. P. and Geller, M. J. (1982) *Astrophys. J.*, **257**, 423.

Kirshner, R. P., Oemler, A., Schechter, P. L., and Shectman, S. A. (1983) *Astron. J.*, **88**, 1285.

Kormendy, J. (1987a) In: *Structure and Dynamics of Elliptical Galaxies, IAU Symposium No. 125*, ed. de Zeeuw, T., Reidel, Dordrecht, The Netherlands.

Kormendy, J. (1987b) In: *Dark Matter in the Universe, IAU Symposium No. 117*, p. 139, eds. Kormendy, J. and Knapp, G. R., Reidel, Dordrecht, The Netherlands.

Kormendy, J. and Knapp, G. R., eds. *Dark Matter in the Universe, IAU Symposium No. 117*, Reidel, Dordrecht, The Netherlands.

Lin, D. N. C. and Lynden-Bell, D. (1982) *Mon. Not. R. Astron. Soc.*, **198**, 707.

Little, B. and Tremaine, S. (1987) *Astrophys. J.*, **320**, 493.

Lynden-Bell, D., Cannon, R. D., and Godwin, P. J. (1983) *Mon. Not. R. Astron. Soc.*, **204**, 87p.

Merritt, D. (1987a) *Astrophys. J.*, **313**, 121.

Merritt, D. (1987b) Submitted to *Astron. J.*

Murai, T. and Fujimoto, M. (1980) *Publ. Astr. Soc. Japan*, **32**, 581.

Primack, J. R. (1987) In: *Proceedings of the International School of Physics "Enrico Fermi"* **92**, ed. Cabibbo, N. Italian Physical Society, Bologna, Italy.

Ostriker, J. P., Peebles, P. J. E., and Yahil, A. (1974) *Astrophys. J. Lett.*, **193**, L1.

Richstone, D. and Tremaine, S. (1986) *Astron. J.*, **92**, 72.

Trimble, V. (1987) *Ann. Rev. Astron. Astrophys.*, in press.

Turner, E. L. (1976) *Astrophys. J.*, **208**, 20.

White, S. D. M., Huchra, J. P., Latham, D., and Davis, M. (1983) *Mon. Not. R. Astron. Soc.*, **203**, 701.

Zwicky, F. (1933) *Helv. Phys. Acta*, **6**, 110.

Chapter 7

GRAVITATIONAL LENSES

Roger D. Blandford* and Christopher S. Kochanek†

Theoretical Astrophysics, California Institute of Technology,
Pasadena, CA 91125, USA

These lecture notes provide an introduction to the theory of gravitational lensing and an assessment of the current observational position. The optics of gravitational lenses is described using vector, scalar, and propagation formalisms and the uses of these three descriptions are outlined. Image properties for isolated lenses of increasing complexity are derived. The importance of including ellipticity and finite core sizes in models of galaxies is emphasized. The properties of singular lenses such as black holes and strings are also distinguished. A topological classification of image combinations is derived using the scalar formalism, and a second classification for highly amplified images together with a general scaling law is obtained using catastrophe theory. Microlensing by stars inside a galaxy can introduce granularity into the image of a point source. The behavior of the microimages is discussed as the optical depth of the stars increases. There is, as yet, no evidence for microimaging. Compound lenses, in which the deflecting mass is localized in two or more deflecting screens, are discussed. Next, a critical review of ten claimed gravitational lens candidates is presented. It is argued that four of these are most probably multiply imaged quasars. One case is almost certainly not a lens and three are probably independent quasars with similar optical spectra. Assessment of the remaining two cases requires further observation. The four strong cases exhibit magnifications (high for bright quasars and of order unity for faint quasars) that reflect the quasar luminosity function. There is, as yet, no need to invoke new forms of dark matter to account for the observed lenses. The prospects for a larger, systematically derived samples of lenses, which is vital for statistical studies, seems reasonably bright. Methods proposed to use gravitational lenses to derive the Hubble constant, the mass density of the universe, and galactic masses are critically analyzed.

1. Introduction

1.1. *History*

The problem in question, however, takes on a radically different aspect, if, instead of in terms of stars we think in terms of *extragalactic nebulae*. Provided that our present estimates of the masses of *cluster nebulae* are correct, the probability

---

*Current address: SLAC, MS 75, Menlo Park, CA 94025, USA.
†Current address: Department of Astronomy, The Ohio State University, Columbus, OH 43210, USA.

that nebulae which act as gravitational lenses will be found becomes practically a *certainty*.

— Zwicky 1937b

The 1919 solar eclipse expedition confirmed the gravitational bending of light by the sun to be the value predicted by general relativity

$$\alpha = \frac{4M_\odot}{r_\odot} = 1.75''$$ (1.1)

at the limb of the sun (in units where $G = c = 1$, which we will use throughout).

This experiment is a classic test of general relativity, and the best experimental data on the bending of light by the sun gives the ratio of the measured to the predicted values to be 1.007 ± 0.009. However, no one *seriously* proposed multiple imaging and amplification until Einstein discussed gravitational lensing by stars in 1936. (Although it was reported to Zwicky that E. B. Frost, a director of the Yerkes observatory, had proposed a search program in 1923.) The first discussion of gravitational lensing by *galaxies* is due to Zwicky in a series of two letters and an article printed in 1937. Even in this series of papers, Zwicky includes the proviso that the masses of the nebulae, and the distances to them, were still a matter of debate. The subject appears to have been ignored until the 1960s when theoretical discussions appeared by Refsdal and Barnothy and Barnothy. The predictions were considered further in the early 1970s by Bourassa, Kantowski, Norton, Press, Gunn, and many others. And so, theorists were not taken by surprise when the first lens 0957 + 561 was discovered in 1979 by Walsh, Carswell, and Weymann.

Since 1979 the field has expanded rapidly. We now have at least ten lens candidates, which are listed in Table 1, with some basic information and our prejudices as to how strong a case exists for each. Theoretical study has likewise burgeoned, with three main topics dominating: models of specific lenses, the statistics of lensing, and microlensing.

1.2. *Simple Estimates*

"The time has come," the Walrus said, "to talk of many things: of shoes — and ships — and sealing wax — Of cabbages — and kings — And why the sea is boiling hot — And whether pigs have wings."

— Lewis Carroll in *Through the Looking Glass*

Given the equation for the bending of light by a spherical mass, we can estimate the requirements for gravitational lenses and the size of the image splittings generated. For a mass M, and impact parameter r, the bending angle

$$\alpha \propto \frac{M}{r} \propto \sigma^2$$ (1.2)

where σ is the characteristic one dimensional velocity dispersion. In particular, for an isothermal sphere, $\alpha = 4\pi\sigma^2 = 2.6''\sigma_{300}^2$. Using the simple geometry shown

Table 1. Lens candidates.

QSO	z_Q	z_L	m_Q	$\theta('')$ arc-seconds	Number of images	How strong a case?	Who
0957 + 561	1.4	0.36	17	6	2	strong	Walsh et al., 1979
1115 + 080	1.7	?	16	2	4		Weymann et al., 1980
2016 + 112	3.3	0.8, ?	22	3	3		Lawrence et al., 1984
2237 + 031	1.7	0.04, 0.6?	17	1	2		Huchra et al., 1985
3C324	1.2	0.84	20	2	2	preliminary	Le Fèvre et al., 1987
1042 + 178	0.9	?	19	3	4		Hewitt et al., 1987
0023 + 171	1.0	?	23	5	2		Hewitt et al., 1987
2345 + 007	2.2	?	19	7	2	controversial	Weedman et al., 1982
1635 + 267	2.0	?	19	4	2		Djorgovski et al., 1984
1146 + 111	1.0	?	18	157	2	not a lens	Turner et al, 1986

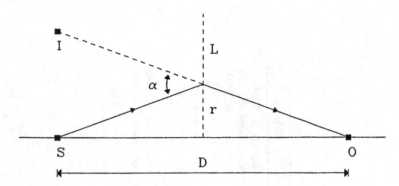

Fig. 1. Schematic diagram of the normal lensing geometry. The observer is at point **O**, the source at point **S**, and the lens is some blob of mass lying in plane **L**. The image **I** is deflected by angle α relative to the source. The impact parameter at the lens is **r**, and **D** is a measure of the distance scale.

in Fig. 1, the condition for multiple imaging is $\alpha D \gtrsim r$ where D is a measure of the distance, and r a measure of the lens size. For cosmological distances, $D \sim (1/3)H_0^{-1} \sim h^{-1}\text{Gpc}$, so that the multiple imaging condition becomes

$$\sigma \gtrsim 100 \left(\frac{r_c}{1\ \text{kpc}} \frac{1h^{-1}Gpc}{D} \right)^{1/2} \text{km s}^{-1} \tag{1.3}$$

where r_c is the core radius of the object defined by the radius of the region in which the surface density falls to half of its peak value. In this and in subsequent equations, the numerical values are more accurate than the arguments used to derive them. Condition (1.3) implies that galaxies generally can create multiple images, but clusters probably cannot. Alternatively, we can deduce a critical surface density of matter Σ for multiple imaging,

$$\Sigma \sim M/r^2 \sim \alpha/r \gtrsim \Sigma_{\text{crit}} \sim 1/D \sim 0.4\ \text{g cm}^{-2} \tag{1.4}$$

for a cosmological distance D. Essentially, as much matter as is found in a column out to $z \simeq 1$ must be compressed into the lens plane. (Multiple imaging can result from arbitrarily small surface densities if the density profile is suitably contrived. However, for reasonable mass distributions, the approximate relation in Eq. (1.4) still stands.) This can also be related to a critical surface brightness,

$$I_0 \lesssim 21 + 2.5 \log \left(\frac{M}{10L} \frac{L_\odot}{M_\odot} \right) \tag{1.5}$$

magnitudes in a square arc second, or gravity, $g \gtrsim 10^{-8}$ cm s^{-2}. Given two such rays, we can estimate the time delay between them to scale as $\Delta t \sim M \sim (1/4)D\theta^2 \sim 0.02(\theta/1'')^2$ years, where θ is the image separation. In principle this provides a means of measuring galactic masses or the Hubble constant.

The probability of lensing can be estimated by an argument due to Press and Gunn. If the lens number density is n, then the optical depth of the universe is approximately $\tau = nr^2D \sim nMD^2 \sim \Omega_{\text{lens}}$. Where Ω_{lens} is the fraction of the critical density to close the universe contained in the lenses. For galaxies this leads to $\tau \sim \Omega_{\text{gal}} \sim 0.01$. While providing a rough estimate, this argument can overestimate the optical depth found using potentials associated with extended mass distributions. Lenses are well described by geometric optics, and are achromatic. Diffraction will be important only if the diffraction length $\sqrt{\lambda D}$ is greater than the typical impact parameter $\sqrt{MD}$ which implies $\lambda \gtrsim M$. Recall that for a one solar mass object, $M \sim 1$ km, so that diffraction will never matter for astrophysical lenses.

1.3. *Uses*

Proposed uses of gravitational lenses include:

(a) **Dark Matter** — Lenses can provide information about dark matter either directly through models of specific lenses and microlensing phenomena, or indirectly through the statistics of lenses. In both cases, only information about clumped dark matter will be unambiguous. While the mean background density contributes to the process of lensing by focusing light rays, the effect is weak compared to the effects of a specific lens.

(b) **Hubble Constant/Galactic Masses** — The time delay between two images of a lens system is a measure of the mass of the lens or the Hubble constant. In either case, the results are highly sensitive to the lens model which is used. This probably means that no single lens will ever yield a value for the Hubble constant which is less subject to systemic error than current methods. The results given by a statistical ensemble of lenses may eventually provide an independent measure of H_0.

(c) **Natural Telescopes** — Gravitational lenses do not make very good telescopes — their behavior is strongly non-linear, and the properties of the cosmological optical bench are largely unknown. Nonetheless, lenses do amplify and magnify distant QSOs. For example, microlensing has the promise of measuring the size of the continuum optical emission region in QSOs on scales far smaller than is possible by any other observational method.

1.4. *Organization of Lectures*

> I mistrust all systematizers and I avoid them. The will to a system is a lack of integrity.
>
> — Nietzsche in *The Twilight of the Idols*

In these lectures, we will try to summarize the field as it stands today. We will proceed by discussing three theoretical approaches to gravitational lensing. These will then be applied to a series of simple lenses to give a feeling for the types of images that ought to be produced in nature. This serves as the beginning of a discussion of the topological properties of image configurations, their stability, and their likelihood. Next we will discuss microlensing, its use and implications for the determination of the size of distant sources, as well as the absence of any observed events. The final theoretical topic will be the complications introduced by having more than one deflection of the light rays by a compound lens. We will discuss the observations next; what is seen, and how it relates to our theoretical knowledge. Finally, we will discuss applications of gravitational lensing to the problem of cosmology and dark matter. In keeping with the lecture format, we have not cited original references in the text. However, we do provide a bibliography at the end of each section which can be consulted for further information on the topics covered.

References

Historical Articles

Barnothy, J. M. and Barnothy, M. F. (1972), "Expected density of gravitational-lens quasars", *Astrophys. J.*, **174**, 477.

Einstein, A. (1936), "Lens-like action of a star by the deviation of light in the gravitational field", *Science*, **84**, 506.

Fomalont, E. B. and Sramek, R. A. (1977), "The deflection of radio waves by the Sun", *Comm. Astrophys.*, **7**, 19.

Refsdal, S. (1964a), "The gravitational lens effect", *Mon. Not. R. Astron. Soc.*, **128**, 295.

Refsdal, S. (1970), "On the propagation of light in universes with inhomogeneous mass distribution", *Astrophys. J.*, **159**, 357.

Zwicky, F. (1937a), "Nebulae as gravitational lenses", *Phys. Rev. Lett.*, **51**, 290.

Zwicky, F. (1937b), "On the probability of detecting nebulae which act as gravitational lenses", *Phys. Rev. Lett.*, **51**, 679.

Zwicky, F. (1937c), *Helv. Phys. Acta*, **6**, 1933.

Review Articles

Canizares, C. R. (1987), "Gravitational lenses as tools in observational cosmology", to appear in *Observational Cosmology*, Proceedings of IAU Symposium 124, Dordrecht: D. Reidel.

Gott, J. R. (1986), "Gravitational lenses and dark matter: Theory", to appear in *Dark Matter in the Universe*, Proceedings of IAU Symposium 124, Dordrecht: D. Reidel.

Narayan, R. (1986), "Gravitational lensing — Models", to appear in *Dark Matter in the Universe*, Proceedings of IAU Symposium 124, Dordrecht: D. Reidel.

Webster, R. and Fitchett, M. (1986), "Cosmology: Views with a gravitational lens", *Nature*, **324**, 617.

2. The Optics of Gravitational Lenses

> For now we see through a glass, darkly;
>
> — I Corinthians 13

There are three separate formalisms that are useful for different problems in gravitational lensing.

2.1. *Vector Formalism*

2.1.1. *The lens equation*

We can derive the equation governing gravitational lensing by the geometrical construction shown in Fig. 2. The lens equation for a single lens follows from simple geometry,

$$\vec{\theta}_S D_{OS} + \vec{\alpha}(\vec{\theta}_I) D_{LS} = \vec{\theta}_I D_{OS}, \qquad (2.1)$$

where $\vec{\theta}_S$ is the angular position of the source, $\vec{\theta}_I$ is the angular position of the image, and the bending angle $\vec{\alpha}$ can be computed by superposing the deflections due to each element of the lens,

$$\vec{\alpha}(\vec{\theta}_I) = 4 D_{OL} \int d^2\theta \frac{\Sigma(\vec{\theta})(\vec{\theta}_I - \vec{\theta})}{|\vec{\theta}_I - \vec{\theta}|^2} \qquad (2.2)$$

where Σ is the surface mass density of the lens.

The distances between the observer and the source D_{OS}, and the lens and source, D_{LS}, are some measure of the distance, the details of which we will discuss later. An equivalent geometrical construction is given by a bending angle diagram, shown in Fig. 3, where the deflection angle is plotted as a function of the image position on the lens plane on which the curve $\theta_I - \theta_S$ is superposed. The images are located at the intersections of the curve and the line. Note that for a non-singular gravitational potential, the deflection angle will be $\propto M/r$ at large radii, and $\propto \Sigma_0 r$ at small radii, where Σ_0 is the central surface density. As long as the potential is non-singular and

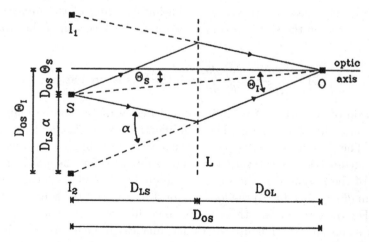

Fig. 2. Diagram of light ray paths in a typical lens. The solid lines show the true path of the light rays from the source **S** to the observer **O** as they are deflected in the lens plane **L**. The observer sees the images I_1 and I_2 of the source as if the light had followed the path of the dashed lines. The lensing equation is shown to be the geometrical construction derived from the distances (D_{OL}, D_{LS}, D_{OS}) and the angles $(\theta_I, \theta_S, \alpha)$ in the problem. The source **S**, which has a real position at θ_S, is seen at the image position θ_I due to a deflection at the lensing mass by angle α.

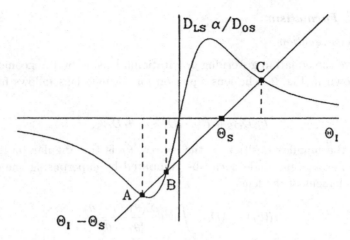

Fig. 3.   A bending angle diagram for a typical spherical lens, characterized by a linear rise at small radii, and a $1/r$ drop off at large radii. To construct the image positions for a source at position θ_S, the line $\theta_I - \theta_S$ is drawn on the bending angle diagram, and the images (A, B, and C) are at the interesections of the bending angle curve and the line.

circularly symmetric, only one or three images will be generated unless the mass distribution is very contrived. Singular potentials will generally capture one image in the singularity of the potential.

2.1.2. *Image amplification and parity*

Since surface brightness is conserved by lensing, the flux of the images depends only on their area on the sky. The amplification is the Jacobian of the transformation (2.1)

$$A = \left| \frac{\partial \vec{\theta}_I}{\partial \vec{\theta}_S} \right| = \frac{\theta_I}{\theta_S} \frac{d\theta_I}{d\theta_S} \quad \text{if the lens is circular.} \tag{2.3}$$

A is the ratio of the areas of an infinitesimal region of the source plane, and its projection onto the image plane. The amplification for a circular lens consists of two parts. The first term, θ_I/θ_S, is the tangential spreading of the image due to the purely geometrical effect that the angle subtended by the image relative to the origin of the potential is the same as the angle subtended by the source. The second term $d\theta_I/d\theta_S$ is due to the focusing or defocusing of the source in the radial direction. For the circular lens this can also be exhibited geometrically as shown in Fig. 4. The same construction also shows that images can be inverted in the process; to see this, label each of the sides of the source and see where that side appears in the images. These changes in the orientation of the images are referred to as their partial parities. The partial parity of an image in any direction is + if it has the same orientation as in the source and − if it has been reflected. Hence each image has a parity label (radial parity, angular parity), and a total parity which is positive

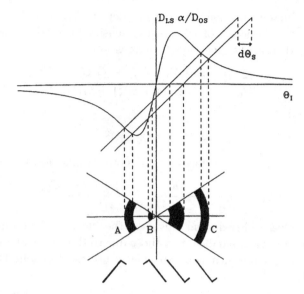

Fig. 4. Construction of image amplifications and parities. For a spherical lens, all images subtend the same angle relative to the origin of the lens as the source does. Hence by taking an extended source from θ_S to $\theta_S + d\theta_S$ and constructing the image positions from the bending angle diagram, the pattern of images on the sky can be found by expanding the images in arcs covering the same angle as the source. The amplification of an image is simply the area of the image on the sky divided by the area of the source. The parities are found by following the mapping of the edges of the source to the edges of the images. Shown at the bottom is the effect of the parities on a bent feature such as a kinked radio jet.

if the partial parities have the same sign and negative if they have opposite signs. The partial parities are the signs of the eigenvalues of the second rank tensor which describes the distortion of the images relative to the source (see Sec. 4.1).

The original image has parity $(+, +)$, and the C image has the same orientation, and hence the same parity, $(+, +)$. The B image has been reflected in both directions, and hence has partial parities $(-, -)$, but the total parity (the product of the two parities) is positive indicating that the orientation can be related to the original orientation by a rotation. The A image is reflected only in the angular direction, giving it partial parities of $(+, -)$ and negative total parity. In practice, the relative (but not the absolute) parities can be established from changes in the orientation of structures in the images — three non-collinear identifiable points are needed, such as can be provided by a kinked radio jet.

2.1.3. *Distance measures*

The universe is, of course, not Euclidean, and the distances used in the lensing equation should not be Euclidean distances. The correct distance measure in a homogeneous universe is the angular diameter distance because it is the distance measure which is defined so as to make the geometric construction described above correct.

In a smooth Friedmann universe with current density fraction $\Omega_0 = 2q_0$ (where q_0 is the deceleration parameter) of the critical density for closing the universe, the angular diameter distance is given by the expression

$$D_{ij} = \frac{2}{H_0} \frac{(1-\Omega_0)(G_i - G_j) + (G_i G_j^2 - G_i^2 G_j)}{\Omega_0^2 (1+z_i)(1+z_j)^2} \tag{2.4}$$

where $z_i < z_j$, and

$$G_i = (1 + \Omega_0 z_i)^{1/2}. \tag{2.5}$$

If point i is the observer, this reduces to the more familiar form

$$D_{Oj} = \frac{2}{H_0} \frac{(1-\Omega_0)(1 - G_j) + (G_j^2 - G_j)}{\Omega_0^2 (1+z_j)^2}. \tag{2.6}$$

If we restrict ourselves to Friedmann universes, the effects of the cosmological model (choice of Ω_0) are generally smaller than those due to the choice of a lens model (see Sec. 8.1). For most purposes, it is acceptable to use the Einstein–De Sitter model in which $\Omega_0 = 1$,

$$D_{ij} = \frac{2}{H_0} \frac{(1+z_i)^{1/2}(1+z_i) - (1+z_i)(1+z_j)^{1/2}}{(1+z_i)(1+z_j)^2} \tag{2.7}$$

and

$$D_{Oj} = \frac{2}{H_0} \frac{(1+z_j) - (1+z_j)^{1/2}}{(1+z_j)^2}. \tag{2.8}$$

The angular diameter distance, and the ratios which commonly appear in the lensing equations are shown in Fig. 5 for sources at redshifts 1, 2, and 3 in an Einstein–De Sitter universe. The major effect is that the universe "stops" beyond some redshift (1 for Einstein–De Sitter) in the sense that the effective deflection angle $\alpha D_{LS}/D_{OS}$ is strongly reduced. For the Einstein–De Sitter case, the most probable lens position is near $z_L = 0.5$ where $D_{LS}/D_{OS} \sim 0.5$ whereas by $z_L = 1.5$, $D_{LS}/D_{OS} \sim 0.1$. This changes slightly when Ω_0 is changed.

If some fixed fraction of the mass density is localized in large clumps, we can still define an approximate angular diameter distance by using the affine parameter and ignoring the matter "outside the beam" (cf. Sec. 2.3).

2.2. Scalar Formalism: Fermat's Principle

With a single non-circular lens, the angles become two dimensional vectors on the sky and it is more convenient to introduce a scalar formalism based on Fermat's principle. In the weak field limit ($\phi \ll 1$), the metric near the lens is

$$ds^2 = -(1 + 2\phi)dt^2 + (1 - 2\phi)dx^i dx_i + O(\phi^2) \tag{2.9}$$

where ϕ is the Newtonian gravitational potential of the lens. The equation for the geodesic followed by a light ray is simply $ds^2 = 0$ which we can integrate along

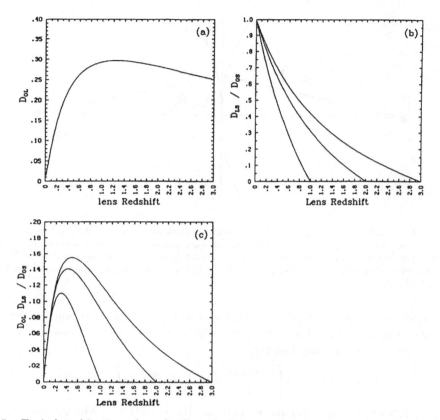

Fig. 5. Typical combinations of angular diameter distances which appear in the lensing equations for an Einstein–De Sitter universe. (a) The angular diameter distance from the observer to a source at a given redshift. (b) The effective bending angle of a lens is $\alpha_e = (D_{LS}/D_{OS})\alpha$ where α is the intrinsic bending angle of the object. Here, D_{LS}/D_{OS} is shown for three source redshifts $z_S = 1, 2$, and 3. (c) The strength of a lens can be measured by the overdensity of the lens relative to a critical surface density $\Sigma_{\mathrm{crit}}^{-1} = 4\pi D_{OL} D_{LS}/D_{OS}$ (see Sec. 3.1). Here we show the cosmological term $D_{OL} D_{LS}/D_{OS}$ for three source redshifts $z_S = 1, 2$, and 3. The factors of D_{OL} and D_{LS} are due to the large surface densities required when the lens approaches either the observer or the source.

the trajectory to determine a time along the path. In a cosmological setting, the lens is always thin and will cause only small angle deflections (except for singular lenses); this allows the use of the "paraxial approximation". After subtracting the time along the geodesic in the absence of the lens we have two contributions to the time delay as measured in the lens plane.

$$t'_{\mathrm{geom}} = \frac{1}{2}\vec{\alpha} \cdot D_{OL}\vec{\theta}_I = \frac{D_{OS}D_{OL}}{2D_{LS}}(\vec{\theta}_I - \vec{\theta}_S)^2, \qquad (2.10\mathrm{a})$$

$$t'_{\mathrm{grav}} = -2\int \phi \, dz. \qquad (2.10\mathrm{b})$$

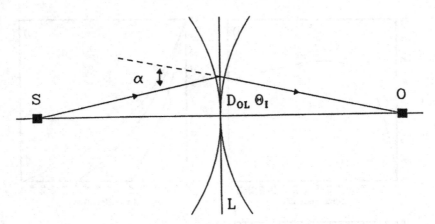

Fig. 6.   Time Delays and Fermat's Principle. Diagram showing the source of the geometrical time delay. The two arcs are contours of constant propagation time from the source and observer. Hence the geometrical time delay is the length of the light path between the two arcs at distance $D_{OL}\theta_I$ from the fiducial ray in the lens plane. Using the fact that the deflection angle is small, the result is simply $\vec{\alpha} \cdot D_{OL}\vec{\theta}_I/2$.

The source of the geometrical contribution is shown in Fig. 6. We must convert to observer time using $t = (1 + z_L)t'$ to give the time delay for a virtual ray with an image at $\vec{\theta}_I$ as seen by the observer,

$$t(\vec{\theta}_I) = (1 + z_L) \left[\frac{1}{2} \frac{D_{OL} D_{OS}}{D_{LS}} (\vec{\theta}_I - \vec{\theta}_S)^2 - 2 \int \phi(\vec{\xi}) dz \right]. \qquad (2.11)$$

Using Fermat's principle, the actual rays are the virtual rays which extremize the time $t(\vec{\theta}_I)$, so that

$$\vec{\theta}_S = \vec{\theta}_I - \frac{D_{LS}}{D_{OS}} 2\vec{\nabla}_{\vec{x}} \phi^{(2)}(\vec{x}) \qquad (2.12)$$

where $\vec{x} = D_{OL}\vec{\theta}_I$. The deflection angle is the gradient of twice the two-dimensional Newtonian potential

$$\phi^{(2)}(\vec{x}) = \phi^{(2)}(D_{OL}\vec{\theta}_I) = 2D_{OL}^2 \int d^2\theta \Sigma(\vec{\theta}) \ln |\vec{\theta}_I - \vec{\theta}|. \qquad (2.13)$$

This is the same result as in Eqs. (2.1) and (2.2) where we identify

$$\vec{\alpha} = 2\vec{\nabla}_{\vec{x}} \phi^{(2)} = 2\vec{\nabla}_{\vec{x}} \int \phi(\vec{x}, z) dz. \qquad (2.14)$$

If we define an effective potential $\psi = 2D_{OL}D_{LS}\phi^{(2)}/D_{OS}$ then the magnification tensor is simply

$$A = \begin{vmatrix} 1 - \partial_x \partial_x \psi & -\partial_x \partial_y \psi \\ -\partial_x \partial_y \psi & 1 - \partial_y \partial_y \psi \end{vmatrix}^{-1}. \qquad (2.15)$$

We discuss the use of the virtual time delay in Sec. 4 below.

2.3. *Propagation Formalism: The Optical Scalar Equations*

The relativistic optical scalar equations were developed by Sachs (1961) and Penrose (1966); they are useful in an inhomogeneous universe where there are no well-defined lens planes at which the light rays are bent. We will give a simple Newtonian discussion. The basic idea is to follow a congruence of light rays from the observer to the source or *vice versa*. If we regard the rays as particles in the plane perpendicular to their direction of motion, we can perform Taylor expansions to find approximate equations for the evolution of the cross section of the congruence. Using the Newtonian equations with the usual doubling of the Newtonian potential to determine the acceleration, the change in the cross section is:

$$\frac{dx_a}{dt} = v_{a,b}x_b = v_a, \tag{2.16a}$$

$$\frac{dv_a}{dt} = -2\phi_{,ab}x_b. \tag{2.16b}$$

The time t is the distance along the ray (for $c = 1$). Equation (2.16) is basically a formula for the tidal distortion of the bundle of rays relative to the path of the fiducial ray (coordinates x_a^f, velocity v_a^f) which is determined by

$$\frac{dx_a^f}{dt} = v_a^f, \tag{2.17a}$$

$$\frac{dv_a^f}{dt} = -2\phi_{,a}. \tag{2.17b}$$

We can rewrite the symmetric tensor $v_{a,b}$ in terms of a real rate of expansion, θ, and a complex shear, $\sigma = \sigma_r + i\sigma_i$,

$$v_{a,b} = \begin{pmatrix} \theta + \sigma_r & \sigma_i \\ \sigma_i & \theta - \sigma_r \end{pmatrix}. \tag{2.18}$$

The symmetry of $v_{a,b}$ implies there is no "angular velocity" and hence no rotation of the images. Note that the acceleration dv_a/dt is just twice the Newtonian gravity in the weak field limit. Combining Eqs. (2.16a) and (2.16b) and using a dot to denote the time derivatives, we get two equations which separately determine the evolution of the beam due to matter inside the beam (Ricci focusing)

$$\dot{\theta} + \theta^2 + |\sigma|^2 = -(\phi_{,11} + \phi_{,22}) = -4\pi\rho \tag{2.19a}$$

and deflections caused by matter outside the beam, (Weyl focusing)

$$\dot{\sigma} + 2\theta_\sigma = -\phi_{,11} + \phi_{,22} - 2i\phi_{,12}. \tag{2.19b}$$

(Equation (2.19b) is really two equations, one for each component of σ.) The matter within the beam focuses the light ray and must decrease the cross sectional area. To see this, consider a bundle of rays with elliptical cross section and area πl^2,

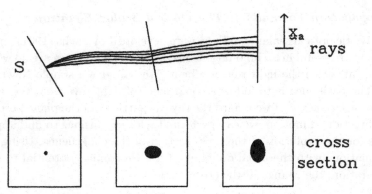

Fig. 7. Optical Scalar Equation. A congruence of rays is show propagating through space, with the variation in the cross section of the bundle about the fiducial ray shown below. The coordinates of the optical scalar equation are measure in the plane perpendicular to the direction of propagation of the fiducial ray.

where l evolves according to

$$\dot{l} = \theta l,$$
$$\ddot{l} = -(\sigma_r^2 + \sigma_i^2 + 4\pi\rho)l. \tag{2.20}$$

The quantity in parenthesis is positive definite as long as the stress tensor satisfies the strong energy condition, and so, this equation implies that the beam will always be focussed (see Fig. 6).

In a cosmological or relativistic situation, the time derivatives become derivatives with respect to an affine parameter $\lambda = \int dt(1+z)^{-1}$ parametrizing the geodesic, and the density and potentials become elements of the Ricci and Weyl tensors. For an Einstein–De Sitter universe ($\Omega_0 = 1; (1+z) \propto t^{-2/3}$) a simple computation gives the affine parameter back to redshift z to be

$$\lambda = 2H_0[1 - (1+z)^{-5/2}]/5.$$

The conclusions are unchanged provided the stress tensor is reasonably well behaved.

References

Vector Formalism

Bourassa, R. R., Kantowski, R., and Norton, T. D. (1973), "The spheroidal gravitational lens", *Astrophys. J.*, **185**, 747.

Bourassa, R. R. and Kantowski, R. (1975), "The theory of transparent gravitational lenses", *Astrophys. J.*, **195**, 13.

Bourassa, R. R. and Kantowski, R. (1976), "Multiple image probabilities for a spheroidal gravitational lens", *Astrophys. J.*, **205**, 674.

Burke, W. L. (1981), "Multiple gravitational imaging by distributed masses", *Astrophys. J. Lett.*, **244**, L1.

Subramanian, K. and Cowling, S. A. (1986), "On local conditions for multiple imaging by bounded smooth gravitational lenses", *Mon. Not. R. Astron. Soc.*, **219**, 333.

Fermat's Principle

Blandford, R. D. and Narayan, R. (1986), "Fermat's principle, caustics, and the classification of gravitational lens images", *Astrophys. J.*, **310**, 568.
Schneider, P. (1984), "A new formulation of gravitational lens theory: Time-delay and Fermat's principle", *Astron. Astrophys.*, **140**, 119.

Optical Scalar Equations

Penrose, R. (1966), in *Perspectives in Geometry and Relativity*, ed. B. Hoffmann (Bloomington: University of Indiana Press).
Sachs, R. (1961), *Proc. Roy. Soc. London*, **264**, 309.

Basic Cosmology and Distance Measures

Schmidt, M. and Green, R. F. (1983), "Quasar evolution derived from the palomar bright quasar survey and other complete quasar surveys", *Astrophys. J.*, **269**, 352.
Weinberg, S. (1972), *Gravitation and Cosmology* (John Wiley and Sons, New York).

3. Gravitational Potential Wells

Let us now examine some simple potentials and their lensing effects. Our procedure is to build up from simple, high symmetry examples to more general potentials that could be used to model a galaxy or cluster.

3.1. *Uniform Sheet*

For a uniform sheet of surface mass density Σ_0, the lensing equation is

$$\theta_S = \theta_I \left(1 - \frac{\Sigma_0}{\Sigma_{\text{crit}}} \right) \tag{3.1}$$

where we define a critical surface density

$$\Sigma_{\text{crit}} \equiv \frac{D_{OS}}{4\pi D_{OL} D_{LS}}. \tag{3.2}$$

The magnification is

$$A = \left(1 - \frac{\Sigma_0}{\Sigma_{\text{crit}}} \right)^{-2} \tag{3.3}$$

which has the property that for $\Sigma_0 = \Sigma_{\text{crit}}$ the sheet focuses the beam onto the observer and the amplification is infinite. Smaller surface densities cannot focus the beam fully, and larger surface densities over focus the beam so that it is diverging again at the observer.

3.2. *Point Mass (Black Hole)*

Let us consider a Newtonian point mass of mass M; two images are always generated as shown in Fig. 8. The lensing equation can be written in the form

$$\vec{\theta}_S = \vec{\theta}_I \left[1 - \frac{\theta_+^2}{|\theta_I|^2} \right]. \tag{3.4}$$

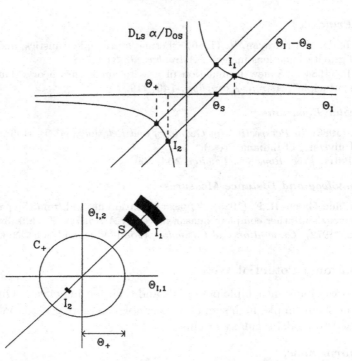

Fig. 8. Bending angle diagram and critical curves of a point mass. The singular lens has one critical line C_+ located at θ_+ in the lens plane corresponding to a degenerate caustic at the origin of the source plane. One image will always appear inside the critical line on the side opposite the source (I_2), and one will appear outside the critical line on the same side as the source (I_1). As the images approach the critical line (source approaching the caustic) they become more tangentially extended, eventually forming an Einstein ring image for a source at the origin. A point mass will always generate two images, but the second image (I_2) becomes increasingly faint as the source moves away from the lens.

It is only necessary to consider radial components of $\vec{\theta}_I$ and $\vec{\theta}_S$. Here $\theta_+ = (2GMD_{LS}/D_{OS})^{1/2}$. Images appear on opposite sides of the center of the potential and are of opposite parity. The amplification A can be expressed as

$$A^{-1} = 1 - \frac{\theta_+^4}{|\theta_I|^4}. \tag{3.5}$$

If the flux ratio of the two images $|A_1/A_2| = R > 1$ so that image 1 is the brighter, then the amplifications of the two images are

$$A_1 = \frac{R}{R-1}, \quad A_2 = -\frac{1}{R-1}. \tag{3.6}$$

If the source is near the optic axis, behind the black hole, there are two images that are very bright and tangentially extended. If the circular symmetry is perfect, an Einstein ring image is formed with radius θ_+ on which the image is infinitely amplified. The curve on the sky formed by the ring image is termed a *critical line*.

The positions of the images are completely determined by R and θ_+,

$$\theta_1 = \theta_+ R^{1/4}, \quad \theta_2 = -\theta_+ R^{-1/4}. \tag{3.7}$$

The brighter image 1 lies outside the critical line, and the fainter image 2 lies inside the critical line.

3.3. *Singular Isothermal Sphere*

The singular isothermal sphere can be used to give a crude model for elliptical galaxies. It is derived from the equations of hydrostatic equilibrium for a spherical distribution,

$$\frac{1}{\rho}\frac{dP}{dr} = g, \tag{3.8a}$$

$$\frac{1}{r^2}\frac{d}{dr}(r^2 g) = -4\pi G\rho, \tag{3.8b}$$

where the isotropic pressure is $P = \rho\sigma^2$. The one-dimensional velocity dispersion σ is taken to be constant, leading to a singular solution with $\rho = (\sigma^2/2\pi G)r^{-2}$. The deflection angle is fixed at all impact parameters, $\alpha = 4\pi\sigma^2 = 2.2''\sigma_{300}^2$ where σ_{300} is the velocity dispersion measured in units of $300\,\mathrm{km\,s^{-1}}$. The lens equation is

$$\vec{\theta}_S = \vec{\theta}_I - \frac{D_{LS}}{D_{OS}}\alpha\frac{\vec{\theta}_I}{|\theta_I|} \tag{3.9}$$

and the amplification is

$$A^{-1} = 1 - \frac{D_{LS}}{D_{OS}}\alpha\frac{1}{|\theta_I|}. \tag{3.10}$$

This results in the bending angle diagram and image configurations shown in Fig. 9. The lens produces two images (of opposite parity) provided

$$\theta_S < \theta_+ = \alpha D_{LS}/D_{OS}.$$

The radius of the C_+ critical line is θ_+, again corresponding to a source on the optic axis. The two images are positioned at $\theta_{1,2} = \theta_S \pm \theta_+$, and the amplifications are $A_{1,2} = \theta_+/\theta_S \pm 1$. We will call this arrangement of the brightest images an *opposed geometry* because the bright images are located on opposite sides of the lens. The condition for the creation of two images is equivalent to having the mean surface density of the lens $\bar{\Sigma}$ inside the critical line greater than the critical density introduced in Sec. 3.1. Again, Σ_{crit} measures the surface density where the amplification becomes infinite, and above which multiple images are created.

3.4. *Isothermal Sphere with Finite Core*

By adding a finite core size and eliminating the singularity at the center of the potential, three images can be created. Let us concentrate on the bright images

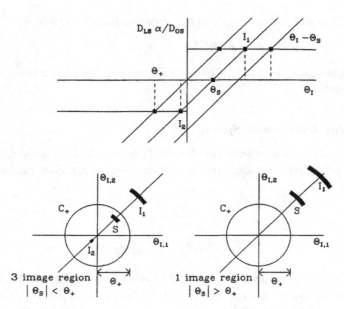

Fig. 9.   Bending angle diagram and critical curves of a singular isothermal sphere. The lens has
only one critical curve C_+ located at θ_+ in the lens plane corresponding to a degenerate caustic
at the origin of the source plane. The amplification is due only to the tangential expansion or
contraction of the image relative to the source.

located near critical lines. The critical lines in the lens plane are images of *caustic lines* in the source plane — hence a source will appear to be strongly amplified
if it is near a caustic line. With a finite core size, there are two critical lines C_+
and C_- located at θ_+ and θ_-. The images near the θ_- critical line are elongated
radially rather than tangentially, and the two bright images straddle the C_- critical
line on the same side of the potential. This arrangement of the brightest images, we
will refer to as an *allied geometry* (see Fig. 10).

In most cases, the third image in the core (opposed geometries) is strongly
deamplified

$$A_3 \simeq \left(\frac{\Sigma_{\text{crit}}}{\Sigma}\right)^2 \sim \frac{r_c^2}{v_{c2}^4} \tag{3.11}$$

where r_c is the core radius in units of kpc and v_{c2} is the central velocity dispersion
in units of $100\,\text{km}\,\text{s}^{-1}$. It is easy for such an image, usually located in the core of
the lens galaxy, to be unobservably faint. In the region inside C_- or outside C_+
the images have positive parity. Between the two critical lines they have negative
parity. In general, the opposed geometry always comprises most of the cross section
for multiple imaging provided the core size is fairly small. If the core size becomes
large, and the lens is only marginally able to generate multiple images (a *marginal
lens*), then the cross section for the allied and opposed geometries are approximately
equal.

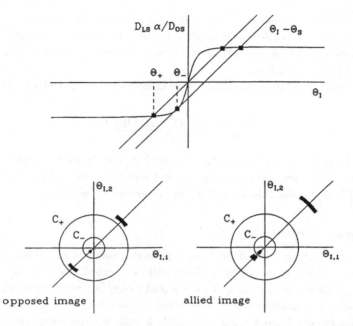

Fig. 10. Bending angle diagram and critical curves of an isothermal sphere with a core. The lens now has two critical curves C_+ and C_- located at θ_+ and θ_- in the lens plane. The C_+ critical line is still due to the degenerate caustic at the origin of the source plane associated with the Einstein ring image. The C_- critical line is located at the point where the line is tangent to the bending angle curve. At this point, the image becomes extended in the radial direction leading to two bright image straddling the C_- critical line. For sources inside the caustic associated with C_-, there are three images, and for sources outside, there is only one image.

3.5. *Elliptical Potentials*

If we introduce some ellipticity into the potential, the lens can generate five images, and the five image cross section exceeds the three image cross section at moderate amplification. Highly amplified images are no longer opposed when the mean amplification of the two brightest images $\bar{A}_{12} \gtrsim \epsilon^{-1}$, where ϵ is the ellipticity of the potential. A simple argument shows that this is a general result. The time delay between two opposed images can be estimated using a Taylor expansion as $\Delta t \sim D\theta_+^2 / \bar{A}_{12}$ where θ_+ is the radius of the C_+ critical line. If the potential has ellipticity ϵ, an azimuthal perturbation $\Delta\phi \sim \epsilon D\theta_+^2$ will be imposed on the arrival time surface near the critical line. When $\Delta t \sim \Delta\phi$, the opposed images will be displaced leading to a five image geometry with the brightest two images allied. Numerical simulations verify this rule for a range of potentials. In general this means that highly amplified images will *not* have large separations on the sky; instead, pairs of bright images will closely straddle critical lines. The total cross section for producing an image with mean magnification of the two brightest images $\bar{A}_{12} > A$ is

$$\sigma(\bar{A}_{12} > A) \sim \frac{S_+}{A^2} \tag{3.12}$$

where S_+ is the area enclosed by the critical lines on the sky. The A^{-2} behavior of the integral cross section is a general result which will be derived in Sec. 4.1. Marginal lenses in which $\Sigma_0 \sim \Sigma_{\text{crit}}$ will generally produce three bright images, the brightest of which are allied.

For all image configurations, the time delay between the two brightest images scales as

$$\Delta t_{12} \sim \frac{(1 + z_L)D_{OL}D_{LS}}{D_{OS}} \frac{\theta_{12}^2}{\bar{A}_{12}} \tag{3.13}$$

where θ_{12} is their separation on the sky and $\bar{A}_{12}$ their mean amplification. This results from Taylor expansions of the lensing equation for image positions near the C_+ critical line. Again, this is verified by numerical simulation.

3.6. *Irregular Potentials*

If several potentials are superposed (as in a small cluster of galaxies), then an irregular pattern of caustics and critical lines will be created (see Fig. 12). There is no simple way to describe the image location and magnification for low amplifications. However, at high amplifications, bright images are found to straddle the critical lines so that results from regular potentials found by expansions near caustics are general. Also as a rule, singularities and near singularities in the potentials will attract and strongly deamplify images associated with the maximum in the arrival time surface.

3.7. *Cosmic Strings*

Cosmic strings result from a cosmological phase transition; the string's metric is determined by the equation of state $P = -\rho$ and it has extent in only one dimension. For an infinite straight string, the metric is

$$ds^2 = -dt^2 + dr^2 + dz^2 + (1 - 4\mu)r^2 d\phi^2 \tag{3.14}$$

where $\mu \sim 3 \times 10^{-6}$ corresponding to a mass per unit length of order kilotons per Fermi. This spacetime is locally flat, but has an angle deficit of $4\pi\mu$ giving a deflection angle

$$\alpha = 4\pi\mu \sin \theta \tag{3.15}$$

where θ is the inclination of the string relative to the optic axis. A moving straight string need not be perpendicular to the line connecting the images. These properties differ from those of line mass in flat space. A straight string produces two images neither of which is amplified, and both of which have the same parity. This allows lensing by a string to be differentiated from other singular lenses, such as a black hole, which produce images of differing parity and amplification (see Fig. 13).

If cosmic strings exist at all, then they are probably irregular and in constant mildly relativistic motion. If they are looped on the scale of the image separation

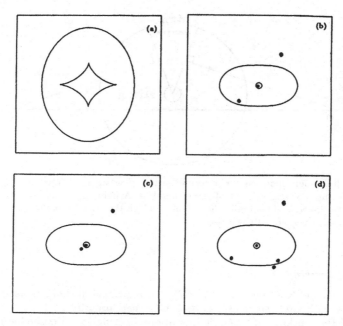

Fig. 11. Caustic lines and critical curves for an elliptical potential. (a) Shows the caustics of an elliptical potential which models a typical galaxy. The inner caustic has four cusps (cf. Sec. 4.2.3) and a source inside it will generate five images. Sources between the outer and inner caustics generate three images, and sources outside both caustics generate only one image. (b) Three image opposed image configuration. The solid lines are the critical curves of the lens, the outer one is C_+ the inner one C_-. Inside C_- and outside C_+ the image parity is even, while between the critical curves, it is odd. (c) Three image allied image configuration. (d). Five image configuration. The brightest two images are allied.

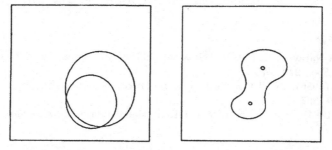

Fig. 12. Caustics of an Irregular Potential. Caustics (left) and critical lines (right) of two super-posed elliptical potentials. In this case only one, three, or five image regions exist. The intersecting caustics form a hyperbolic-umbilic catastrophe (cf. Sec. 4.2.5).

then the metric will change on the scale of the time it takes light to pass the string (because the internal oscillations of the string are strongly relativistic). In this case, Fermat's principle no longer applies, and there is essentially no restriction on image type and behavior.

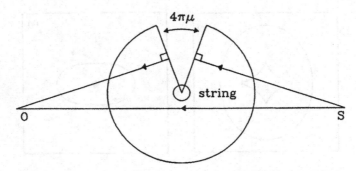

Fig. 13.   Bending angle diagram for a cosmic string. The wedge shaped region missing from the circle represents the angle deficit of the string $4\pi\mu$. A light ray enters and leaves the wedge perpendicular to the edge. This limits the region in which the string generates a second image to an area of $4\pi\mu$ on either side of the string.

References

Examples of Lensing

Blandford, R. D. and Kochanek, C. S. (1987), "Gravitational imaging by isolated elliptical potential wells I: Cross sections", *Caltech GRP-086*, preprint.
Katz, N. and Paczyński, B. (1987), "Gravitational lensing by an ensemble of isothermal galaxies", *Astrophys. J.*, **317**, 11.
Kovner, I. (1987), The quadrupole gravitational lens", *Astrophys. J.*, **312**, 22.
Kovner, I. (1987), The marginal gravitational lens", *Astrophys. J.*, in press.
Narasimha, D., Subramanian, K., and Chitre, S. M. (1986), "Missing image' in gravitational lens systems?", *Nature*, **321**, 45.
Narayan, R., Blandford, R. D., and Nityananda, R. (1984), "Multiple imaging of quasars by galaxies and clusters", *Nature*, **310**, 112.
Schneider, P. and Weiss, A. (1986), "The two point mass lens: Detailed investigation of a special asymmetric gravitational lens", *Astron. Astrophys.*, **164**, 237.

Cosmic Strings

Hogan, C. and Narayan, R. (1984), "Gravitational lensing by cosmic strings", *Mon. Not. R. Astron. Soc.*, **211**, 575.
Paczyński, B. (1986), "Will Cosmic Strings be Discovered Using the Space Telescope?", *Nature*, **319**, 567.
Vilenkin, A. (1984), "Cosmic Strings as Gravitational Lenses", *Astrophys. J. Lett.*, **282**, L51.

4. Generic Features of Images

4.1. Arrival Time Surfaces

We next want to examine the properties of the arrival time surface formed by the virtual time delay introduced in Sec. 2.2. We define the renormalized time delay

$$r = \frac{D_{LS}t}{(1 + z_L)D_{OL}D_{OS}} \tag{4.1}$$

and the renormalized two dimensional Newtonian potential (cf. Sec. 2.2)

$$\psi = \frac{2D_{OL}D_{LS}}{D_{OS}}\phi^{(2)} = \frac{4D_{OL}D_{LS}}{D_{OS}}\int d^2\theta'\Sigma(\vec{\theta}')\ln(\vec{\theta}_I - \vec{\theta}'). \tag{4.2}$$

The potential ψ satisfies the Poisson equation in two dimensions with the source equal to twice the density measured in units of the critical density (cf. 3.1) for lensing

$$\nabla^2\psi = \frac{2\Sigma}{\Sigma_{\text{crit}}}. \tag{4.3}$$

Expressed in terms of ψ, the time delay is

$$r = \frac{1}{2}(\vec{\theta}_I - \vec{\theta}_S)^2 - \psi(\vec{\theta}_I). \tag{4.4}$$

For the moment, we can choose our image plane coordinates to be centered on the source position $\vec{\theta}_S$ so that

$$r = \frac{1}{2}\vec{\theta}_I^2 - \psi(\vec{\theta}_I). \tag{4.5}$$

The stationary points of the virtual time correspond to extrema in the time delay surface so that the location of images is easily determined by examining a contour plot of the time delay. If $\psi = 0$ the surface is a parabola centered on $\vec{\theta}_I = \vec{0}$ and the image is located at the origin. If $\psi \neq 0$ the contours become distorted (see Fig. 14). In particular, contours which pass through saddle points, which we will call *crossing contours*, give a topological classification of allowed image geometries. The time delay can also be used directly for attempts to determine the Hubble constant or the mass of the lens from observations, when it is evaluated along the actual ray.

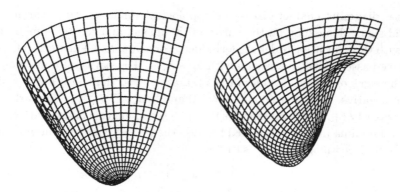

Fig. 14.   Time Delay Surfaces. The figure on the left is of the unperturbed parabolic time delay surface. The figure on the right shows the surface when it is sufficiently perturbed to multiply image. The crossing contour which characterizes this time delay surface is the limaçon.

The magnification of an image (cf. Sec. 2.2) is determined by the extrinsic curvature of the surface

$$\tau_{,ij} = \frac{\partial \theta_{Si}}{\partial \theta_{Ij}}. \tag{4.6}$$

Physically, $\tau_{,ij}$, which we will call the magnification tensor, measures the ratios of the areas of a small region on the source plane, and its projection onto the image plane. The determinant $H = |\tau_{,ij}|$, which is called the Hessian of the transformation $\vec{\theta}_I \to \vec{\theta}_S$, is the inverse of the image amplification, $H^{-1} = A$.

Notice that if the time surface is locally flat (small curvature) the image will be strongly amplified. The magnification tensor is symmetric, so its diagonalized form can be written as

$$\tau_{,ij} = \begin{pmatrix} \kappa + \mu & 0 \\ 0 & \kappa - \mu \end{pmatrix} \tag{4.7}$$

where $\kappa = 1 - \Sigma/\Sigma_{\text{crit}}$ and μ are respectively the expansion and the shear of the image. If we imagine increasing the lens distance D_{OL} then the rates of change of κ and μ are related to the quantities θ and σ introduced in Sec. 2.3. The signs of the two eigenvalues $\kappa + \mu$ and $\kappa - \mu$ determine the parity of the images, and the character of the extremum in the time delay at which the image is located. If the signs of the eigenvalues are $(+, +)$ the extremum is a minimum, if they are $(+, -)$ the extremum is a saddle point, and if they are $(-, -)$ the extremum is a maximum. In general, if there are N images, $(N + 1)/2$ will have positive total parity, and $(N - 1)/2$ will have negative total parity. High resolution observations of gravitational lenses with the HST (or VLBI if the images have radio structure) allow the determination of the relative parities of the images. Because the Hessian is formed from the second derivatives of a scalar function, the magnification tensor is symmetric, and the images can be sheared or reflected, but not rotated. This is no longer true if there is more than one lens plane contributing to the deflection (Sec. 6).

These diagrams characterize some topological limits on time variations and potential shape depending on the parity of the image, and certain combinations of parity with time variation order are forbidden. For example, all maxima (cf. Sec. 2.3) must have passed through a region with $\Sigma > \Sigma_{\text{crit}}$. The parity of a maximum, and hence the signs of the eigenvalues of (4.7), are $(-, -)$ so that both $\kappa + \mu$ and $\kappa - \mu$ must be negative. This is true only if $\kappa < 0$ and hence $\Sigma > \Sigma_{\text{crit}}$. Similarly, a minimum has parity $(+, +)$ which requires that $\kappa > 0$ and hence a region with $\Sigma < \Sigma_{\text{crit}}$. In fact, all minima must be amplified because the $(+, +)$ parity requirement implies $|\mu| < \kappa < 1$ so the magnification satisfies

$$A = \left(\kappa^2 - \mu^2\right)^{-2} > 1.$$

The first image to vary must be a minimum because the time delay surface approaches $+\infty$ for large $|\theta_I|$. This means the time delay surface must have a global minimum. This result is general and will hold for an arbitrary odd number

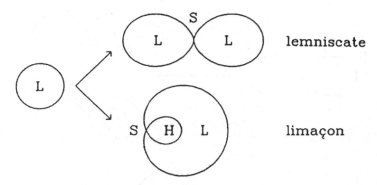

Fig. 15. Crossing contours for three image topologies.

of images, even with multiple lens planes. Near the transitions between topologies, there are strongly amplified images. At least one eigenvalue of the curvature tensor passes through zero as the critical curve is crossed. Hence, the magnification $|\tau_{,ij}|^{-1}$ becomes infinite.

By examining the crossing contours of the time delay, we can classify the allowed parity combinations if the time delay asymptotically becomes parabolic. The "odd image theorem" which states that non-singular lenses generate only odd numbers of images also becomes geometrically clear when lensing is viewed in terms of deformations of a parabolic surface. When there is only one image, the only extremum is a minimum and there are no crossing contours of the surface (see Fig. 15). If there are three images, then there are two topologically distinct cases. The first is called a *limaçon*, and the second is called a *lemniscate*. The limaçon has a minimum, a maximum, and a saddle point, and its limiting case is a black hole which generates two images of different parity. The lemniscate has two minima and a saddle point, and its limiting case, the cosmic string, generates two images of the same total and partial parities (cf. Sec. 3.2). If we allow five images, there are six different non-degenerate geometries (see Fig. 16). For seven images, there are 25 different allowed non-degenerate crossing contours.

4.2. *Caustics and Catastrophes*

4.2.1. *Structural stability of images*

Caustics are the loci of source positions for which two or more associated image magnifications diverge to infinity. (Normally in optics we imagine the caustics as existing in the image space of a fixed point source. However, in the application to gravitational lenses, it is more convenient to fix the observer and construct the complementary caustics in the source space.) The nature of the caustics is sensitive to the degree of symmetry in the lens. If there is sufficient asymmetry, then the caustics will be *structurally stable* to small perturbations in the lens. However, high symmetry (e.g. *circularly symmetric* lenses) produce *structurally unstable* caustics

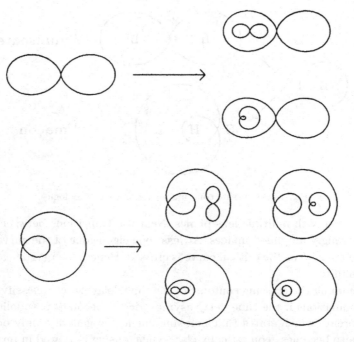

Fig. 16.   Crossing contours for five image topologies.

which change their characteristics as soon as the symmetry is broken. The caustics can be described by catastrophe theory, which has a long tradition of application to optics (albeit not under this name), and a shorter history of imaginative application to the social sciences.

4.2.2. *Isolated image*

Before considering the form of catastrophes, let us discuss the behavior of the time delay surface for an isolated image. If we choose the origin of the lens and source spaces so that a source located at $\theta_S = 0$ produces and image at $\theta_I = 0$, then the time delay surface is locally

$$r(\theta_1) = \frac{1}{2}ax_I^2 + \frac{1}{2}by_I^2 + \cdots \tag{4.8}$$

where x_I and y_I are Cartesian coordinates and an additive constant has been dropped. The constants a and b are derived from a Taylor expansion of the Newtonian potential about the image point, and $a+b \leq 2$ is required for a positive surface density. For a general source position θ_S the images are located at

$$x_I = \frac{1}{a}x_S, \quad y_I = \frac{1}{b}y_S, \tag{4.9}$$

which means the mapping is simply a stretching of the coordinate systems by factors of $1/a$ and $1/b$ along the coordinate axis. A circular source is mapped into an ellipse, with magnification $A = (ab)^{-1}$ and axis ratio b/a. This fully characterizes the *local* behavior of an isolated image.

4.2.3. *Fold catastrophe*

The simplest catastrophe is the fold catastrophe, which corresponds to caustics tracing a smooth curve in the source plane. A source on one side of a fold generates two images which straddle a critical line in the lens plane, while on the other side it generates no images (see Fig. 17). As the source crosses the fold, the two images merge producing strongly amplified images. Locally, the time delay surface can be written

$$r(\theta_I) = \frac{1}{3}ax_I^3 + \frac{1}{2}bx_I^2 + \sqrt{bc}x_I y_I + \frac{1}{2}cy_I^2 \qquad (4.10)$$

where $a > 0$, without loss of generality. The image positions are

$$x_I = \pm\frac{1}{\sqrt{a}}\left[x_S - \sqrt{\frac{b}{c}}y_S\right]^{1/2} \quad , \quad y_I = \frac{1}{c}y_S - \sqrt{\frac{b}{c}}x_I \qquad (4.11)$$

and the magnification of each image is

$$A = \frac{1}{2acx_I} = \frac{1}{2c\sqrt{a}(x_S - \sqrt{b/c}y_S)^{1/2}}. \qquad (4.12)$$

For $x_S > (b/c)^{1/2}y_S$ there are two images, and for $x_S < (b/c)^{1/2}y_S$ there are no images. When $x_S = (b/c)^{1/2}y_S$ the two images merge at $x_I = 0$ with infinite amplification. Thus, the line $x_I = 0$ is a critical line, and $x_S = \sqrt{b/c}\,y_S$ is a caustic. If a source approaches a fold at x_0 with constant velocity, the amplification of the image $A \propto |x-x_0|^{-1/2} \propto |t-t_0|^{-1/2}$ where t_0 is the time at which the source reaches the fold. Globally, there must be another image to satisfy the odd-number theorem

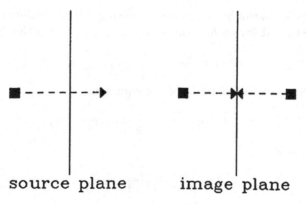

source plane   image plane

Fig. 17. Fold Catastrophe. Diagrams of the source plane and image plane near a fold as a source crosses over the fold starting from the multiple imaging side. The images start on opposite sides of the fold, and merge when the source is on the fold caustic where they are infinitely amplified. When the source has crossed the catastrophe there are no images (locally, globally there must be at least one if the lens is non-singular).

(assuming a non-singular lens). To this order, a fold caustic is locally a straight line; globally, the fold must form a closed loop. If the caustic has self-intersections, these may correspond to higher order catastrophes. As the images appear on opposite sides of the critical line, they will have opposite parities.

These properties, and the fact that essentially all of the length of the caustics in a lens consist of folds, allows a simple calculation of the cross section of a lens as a function of amplification. The cross section is

$$\sigma = \int d^2\theta_S = \frac{1}{2}\int d^2\theta_I H(\theta_I) \tag{4.13}$$

where $H(\theta_I)$ is the Hessian (cf. Eq. (4.6)). The factor of $1/2$ is needed to prevent double counting, as there are two images per source near a fold catastrophe. If we perform a Taylor expansion of the Hessian near a critical line, decomposing the integral into directions locally parallel ($\theta_{||}$) and perpendicular ($\theta_\perp$) to the critical line

$$\sigma \simeq \frac{1}{2}\int d\theta_{||}d\theta_\perp \frac{\partial H}{\partial \theta_\perp}\theta_\perp$$

$$\simeq \frac{1}{2A^2}\int \frac{d\theta_{||}}{|\nabla_{\theta_\perp}H|_{H=0}}. \tag{4.14}$$

This gives the general rule that the cross section for amplification greater than some value A is proportional to A^{-2} if the caustic structure is dominated by folds. This result is verified by numerical calculations. Hence there is a low probability for amplifications of $A \gg 1$, but there are many more faint quasars than bright quasars and it may be possible that the two effects balance (cf. Sec. 7.4).

4.2.4. *Cusp catastrophe*

The next generic catastrophe is the cusp, which generates either three images or one image. The time delay surface can be expanded locally in the form

$$r(\theta_I) = \frac{1}{4}ax_I^4 + \frac{1}{2}bx_I^2 y_I + \frac{1}{2}cy_I^2 \tag{4.15}$$

with images located at the solutions of the equations

$$ax_I^3 + bx_I y_I = x_S, \quad \frac{1}{2}bx_I^2 + cy_I = y_S. \tag{4.16}$$

The magnification is

$$A = \frac{1}{(3ac - b^2)x_I^2 = bcy_I}. \tag{4.17}$$

The critical line and the caustic are the curves for which the magnification diverges: the critical line is a parabola,

$$y_I = \frac{b^2 - 3ac}{bc}x_I^2 \tag{4.18}$$

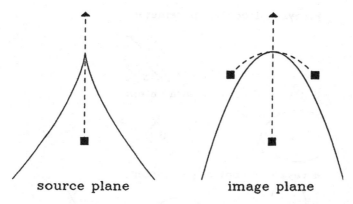

source plane   **image plane**

Fig. 18.   Cusp Catastrophe. Diagrams of the source plane and image plane near a cusp as a source crosses through the cusp. There are initially three images, which merge as the source approaches the cusp. On the cusp there is one infinitely amplified image, and when the source has passed the cusp, there is only one image.

and the caustic is

$$y_S^3 = \frac{27c^2}{8b} \left(1 - \frac{2ac}{b^2}\right) x_S^2. \tag{4.19}$$

For sources "inside" the caustic there are three images, and for sources "outside" there is one image. When $\theta_S = 0$ the three images merge at $\theta_I = 0$ leaving one infinitely amplified image at the same location. As the caustic is approached, the images become elongated parallel to the critical line (see Fig. 18). In general, cusps consist of only a tiny fraction of the total length of caustics in a lens, and hence do not usually affect the scaling laws of Sec. 4.2.4 (for an exception, see Sec. 5.4).

4.2.5. *Higher order catastrophes*

Higher order catastrophes will occur in non-degenerate lenses although the cross section associated with them will be much lower than for the fold. In general, as we vary the observer's position the number of images changes, but there are relatively few ways this can happen in a generic case. What we need is a classification of the way crossing contours can have merging extrema (e.g. source position coordinates or lens parameters). Suppose we have n rays involved in a catastrophe; then we need $(n-1)$ "control" parameters to explore the general case, measuring the angles between images. The image positions θ_I are called "state" variables.

For example, with only two rays ($n = 2$), the arrival time surface must look like Fig. 19 with only one control parameter, the separation between the images. This corresponds to a fold catastrophe. With three rays ($n = 3$), there are two control parameters, and we can form cusps. Notice that this can occur for the lemniscate

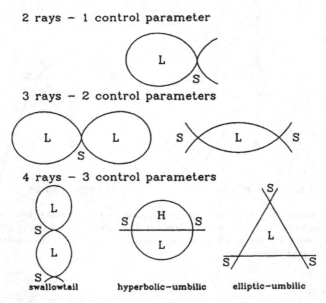

Fig. 19. Higher order catastrophes. Schematic drawings of the crossing contours of higher order catastrophes, for the case in which the saddle points all share the same contour. The general case, when the saddles do not share the same contour, can be simply derived from these. Similar diagrams can be drawn for five ray catastrophes.

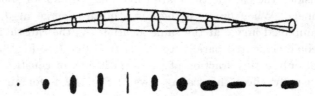

Fig. 20. Optical scalar equation and caustics. The upper diagram shows the bundle propagating through space. The lower diagram shows the cross section of the bundle. Caustics occur when the cross section becomes degenerate and forms a line.

topology, but not the limaçon topology. If there are more than two control parameters, a little experiment suffices to determine the topologically distinct nestings of crossing contours. With $n = 4$ we can form three distinct catastrophes, which are known as the swallowtail, the hyperbolic umbilic, and the elliptic umbilic. Finally, with $n = 5$ there are the butterfly and parabolic umbilic catastrophes. For a single elliptical lens, the highest order generic catastrophe is the hyperbolic umbilic (see Figure 12).

4.3. Caustics as Conjugate Points of Ray Congruences

The catastrophes can be interpreted from the point of view of the optical scalar equation in terms of congruences of rays (see Fig. 20). If we follow a pencil of rays with circular cross section about the fiducial ray near the source, than this cross section will be distorted along the ray, and will in general be elliptical. Eventually, the cross section may degenerate into a line where the rays cross the fiducial ray. The point where this occurs is conjugate to the source, and observer located there would clearly see a point subtend a finite solid angle and so be infinitely magnified. At this point, the source lies on a caustic. If the congruence is traced further, it may yield additional conjugate points. If we now trace the rays backwards in time from the observer, we can trace out the source caustics as the loci of points at the source which are conjugate to the observer. The caustic surfaces will generally be smooth on the scale of the scattering potential.

References

Caustics and Catastrophes

Arnold, V. I. (1984) *Catastrophe Theory* (Springer-Verlag, New York).

Berry, M. V. and Upstill, C. (1976) *Progress in Optics XVIII*, **2**, 57.

Poston, T. and Stewart, I. N. (1978) *Catastrophe Theory and Its Applications* (Pitman, London).

Thom, R. (1975) *Structural Stability and Morphogenesis* (Benjamin, Reading, Mass.).

5. Microlensing

5.1. *Order of Magnitude Estimates*

Using the order of magnitude estimates for the scale of angular deflections, we estimate the mean deflection for objects of mass M to be

$$\theta \sim (M/D)^{1/2} \sim (M/M_\odot)^{1/2} \mu\text{as} \tag{5.1}$$

where D is the effective cosmological distance. The optical depth for the occurence of microlensing can be estimated from the angular cross section for microlensing, $\sigma \sim \pi\theta^2$, and the number of stars per unit angular area n_*. The optical depth is

$$\tau \sim f_* \Sigma/\Sigma_{\text{crit}} \tag{5.2}$$

where f_* is the fraction of the local mass in stars, and Σ is the local surface density of the galaxy. In the outer parts of galaxies, $\Sigma \ll \Sigma_{\text{crit}}$ and $f_* \ll 1$ so that galaxies are optically thin; but in the central regions, $\Sigma > \Sigma_{\text{crit}}$ and $f_* \sim 1$ so that the optical depth exceeds unity and the problem becomes non-linear due to multiple scatterings. If we model the galaxy by an isothermal sphere, the central surface density is

$$\Sigma_0 = \frac{9\sigma_0^2}{2\pi r_c} \cong \frac{6\sigma_{0,300}^2}{r_{c,kpc}} \text{ g cm}^{-2}.$$

Recall from Eq. (1.4) $\Sigma_{\text{crit}} \sim 1 \, \text{g cm}^{-2}$, so that the optical depth in the center of galaxies is $\tau \sim 10$. The impact parameter with respect to the star is $b_* \sim \sqrt{MD} \sim$

$10^{16}(M/M_\odot)^{1/2}$ cm $\gg r_*$ where r_* is the radius of the star. This condition allows the microlensing potential to be treated as that of a point mass.

Another phenomenon which we might call "millilensing" can be caused by objects whose masses lie between stellar and galactic masses such as globular clusters, giant molecular clouds (GMCs), and hypothetical halo black holes with masses of order $10^6 M_\odot$. In this range of masses, the characteristic scale of the deflection angle is milliarcseconds. They will only produce multiple images if they have surface densities $\Sigma > \Sigma_{crit}$. Globular clusters certainly satisfy this condition, but the fraction of galactic mass in these objects is small, so the optical depth will be small and globular cluster lensing will be correspondingly rare. Giant molecular clouds have the opposite problem, because although they contain a large fraction of the mass, they are not dense enough to cause multiple imaging. Local condensations within molecular clouds, however, may exceed the critical surface density. The dark mass in galactic halos may be contained in black holes with masses $\lesssim 10^6 M_\odot$: heavier objects will fall into the galactic center through the effects of dynamical friction. Limits on the incidence of millilensing will eventually set limits on the mass density of objects in this mass range.

If the angular size of the source is much larger than the scale of the deflections then microlensing becomes unimportant. When the lensed objects are quasars, we can estimate whether or not the source size scales are small enough for microlensing to occur. Quasar radio and optical emission lines come from regions with scales $\sim 3\,\mathrm{pc} \gg 10^{16}$ cm, which means that microlensing will not occur, but "millilensing" might. The size of the optical continuum source is unknown. If the UV bump is roughly black body radiation from the inner regions of an accretion disk, then the size is 10^{15} to 10^{16} cm, which would allow microlensing to occur. However, time variability on the scale of months to years implies a source size of $\sim 10^{17}$ cm which preclude it. The X-ray flux can vary on scales of weeks to days, which may imply an emission region small enough for microlensing to be important.

The predicted time scales for the variations due to microlensing range from weeks to centuries depending on the impact parameter and the relative velocities of the source and microlensing object on the sky. The time scale must be roughly

$$t \sim b_*/v = 30(M/M_\odot)^{1/2} v_{100}^{-1}$$

years where v_{100} is the relative velocity in units of $100\,\mathrm{km\,s^{-1}}$. The source of the velocity can be the velocity dispersion of the stars, the motion of the source, if the microlens' galaxy is in a cluster the motion of the galaxy, and the motion of the observer. The relative velocity may be up to $v \sim 1000\,\mathrm{km\,s^{-1}}$ in which case $t \sim 3$ years. Canizares has compared the ratio of continuum emission to line emission since microlensing should change the relative profiles. No evidence for such behavior has been found, implying that microlensing is generally unimportant for quasars.

5.2. *The Character of Microimages: Low Optical Depth*

At low optical depths, we can consider one star at a time, and the star can be modelled as a point mass. The character of the microlensed images depends on what the time delay surface of the macro image would be in the absence of microlensing.

If the unperturbed image is at a minimum then the local time delay surface is deformed into a limaçon with the maximum located on the star. Thus there are two images, one at the minimum, and one at the saddle point. For simplicity, assume there is no shear so the time delay is locally given by

$$r = \frac{1}{2}a\theta_I^2 - \frac{4M}{D}\ln|\theta_I - \theta_*| \tag{5.3}$$

in coordinates θ_I centered on the position of the unperturbed image, where $D = D_{OS}D_{OS}/D_{LS}$, $a \equiv 1 - \Sigma_{\text{smooth}}/\Sigma_{\text{crit}}$, and θ_* is the distance from the star to the unperturbed image. The images are located at the solutions of

$$a\theta_I = \frac{4M}{D}\frac{1}{\theta_I - \theta_*}. \tag{5.4}$$

If we define a critical angle $\theta_c = (4M/aD)^{1/2}$, then the minimum is located at $\theta_I \simeq -\theta_c(\theta_c/\theta_*)$ with amplification a^{-2}, and the saddle is located at $\theta_I \sim \theta_* + \theta_c(\theta_c/\theta_*)$ with amplification $(\theta_c/\theta_*)^4 \lesssim 1$ provided $\theta_* \gg \theta_c$. The image at the saddle will be faint, and if there are many uncorrelated microlenses there will be a bright image plus many faint images with amplification $A \propto (\theta_c/\theta)^4$ the bright image.

If the unperturbed image is at a maximum maximum (this is the most likely case for multiply imaged sources since maxima are usually associated with galactic cores where the optical depth is highest), we can use the same theory, except $\Sigma_{\text{smooth}} > \Sigma_{\text{crit}}$ so that $a < 0$. The local time delay surface will be a lemniscate with two maxima and a saddle point, the second maximum being absorbed by the star. However, if $\theta_* \leq \theta_c$ then there is only one solution to the lensing equation. In this case the saddle merges with the original maximum, annihilating it and leaving only the maximum pinned inside the star. Effectively, the star absorbs the original image leaving an even number of images globally — this is one possible solution to the odd image problem.

If the unperturbed image is a saddle point, then two images will be created. This may apply to the B image in 0957 + 561 which is only $1''$ from the galactic core. However, in this case the approximation of circular symmetry used for maxima and minima is suspect.

5.3. *The Character of Microimages: Moderate Optical Depth*

The optical depth becomes moderately large if the number of stars within θ_c of another star is approximately equal to one. This requires $\Sigma_*/a\Sigma_{\text{crit}} \sim 1$ so that several stars interact. Two new features develop: the first is that the shear from other nearby stars amplifies the microlensed images, and the second is that, rather

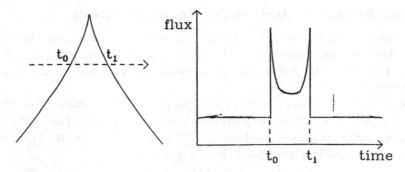

Fig. 21.   Amplification signature of microlensing event. After the source crosses the fold at t_0 (at which time the amplification is formally infinite), the observed flux decays as $|t - t_0|^{-1/2}$, and similarly the flux rises again as the source approaches the second caustic at time t_1.

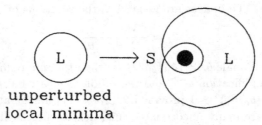

unperturbed
local minima

Fig. 22.   Crossing contours for microlensing at a minimum. The unperturbed minimum is distorted into a limaçon crossing contour. The black region covers the maximum, which is unobservable because it is inside the star.

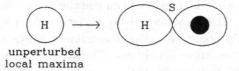

unperturbed
local maxima

Fig. 23.   Crossing contours for microlensing at a maximum. The unperturbed maximum is distorted into a lemniscate crossing contour with two maxima and a saddle. The black region covers one of the maxima, which is unobservable because it is inside the star.

than isolated caustics, a caustic network begins to develop as the caustics from one star overlap with another. As an image crosses a fold of the caustic network, a cusp will appear in the source intensity which varies as $t^{-1/2}$ where t is the time away from crossing the fold. The resolution of the intensity cusp depends on the source size, and microlensing could be a good way of determining source sizes on scales far smaller than any other observational method. For multiply imaged systems, intrinsic source fluctuations can be distinguished from microlensing events by comparing the light curve of the microlensed image with the light curves at the other images. In principle parallax observations within the solar system could be used to separate microlensing effects from intrinsic source fluctuations for singly imaged objects.

We can crudely regard each star as having a fold length associated with its caustics of $l_F \sim 2\pi\theta_c$, and, given the local density of stars, the fold spacing is roughly

$$\Delta_{\text{fold}} \sim \frac{M_*}{D_{OL}^2 \Sigma_* l_F} \sim \frac{\Sigma_{\text{crit}}\theta_c}{2a\Sigma_*}. \qquad (5.5)$$

If $\Sigma_* \gg \Sigma_{\text{crit}}$ then the interval between fold crossings becomes short, and fluctuations occur on time scales much shorter than 30 years. The peak magnification of the image depends on the angular size of the source $\Delta\theta$, with

$$A_{\text{max}} \sim \left(\frac{\Delta\theta}{\theta_c}\right)^{-1/2} \sim 3\left(\frac{R_{\text{source}}}{10^{15} \text{ cm}}\right)^{-1/2}. \qquad (5.6)$$

The most optimistic assumptions give $\Delta\theta/\theta_c \sim 0.01$ in which case $A_{\text{max}} \sim 10$, and pessimistic assumptions give $\Delta\theta/\theta_c \sim 1$ in which case $A \sim 1$ and microlensing is difficult to recognize. The estimated size of optical continuum emission region in quasars is large enough that we are closer to the latter case than the former.

5.4. *The Character of Microimages: Large Optical Depth*

At large optical depths, the idea of a caustic network becomes less useful. Folds may no longer completely dominate the cross sections so that the A^{-2} cross section law may be violated. The variations become much smoother and it is difficult to separate intrinsic variations from the effects of microlensing. The overall effect may be described by a point spread function, with a central Gaussian core and a power law tail $\propto \theta^{-4}$ where θ is the distance from the central intensity peak. This is in reasonable agreement with numerical simulations.

References

Chang, K. and Refsdal, S. (1979) "Flux variations of QSO 0957 + 561 A, B and image splitting by stars near the light path", *Nature*, **282**, 561.

Chang, K. and Refsdal, S. (1984) "Star disturbances in gravitational lens galaxies", *Astron. Astrophys.*, **132**, 168.

Deguchi, S. and Watson, W. D. (1986), "Electron scintillation in gravitationally lensed images of astrophysical radio sources", University of Illinois (Urbana-Champaign), preprint.

Grieger, B., Kayser, R., and Refsdal, S. (1986) "A parallax effect due to gravitational micro-lensing", Hamburger Sternwarte, preprint.

Kayser, R., Refsdal, S., and Stabell, R. (1986) "Astrophysical implications of gravitational micro-lensing", Hamburger Sternwarte, preprint.

Kayser, R., Refsdal, S., Stabell, R., and Grieger, B. (1987) "Astrophysical applications of gravitational micro-lensing", to appear in *Observational Cosmology*, Proceedings of IAU Symposium 124, Dordrecht: D. Reidel.

Nemiroff, R. J. (1986) "Prediction and Analysis of Basic Microlensing Phenomena," PhD Thesis, University of Pennsylvania.

Nityananda, R. and Ostriker, J. P. (1984) "Gravitational lensing by stars in a galaxy halo: Theory of combined weak and strong scattering", *J. Astrophys. Astron.*, **5**, 235.

Ostriker, J. P. and Vietri, M. (1984) "Are some BL lacs artifacts of gravitational lensing?", *Nature*, **318**, 446.

Paczyński, B. (1986) "Gravitational micro-lensing at large optical depth", *Astrophys. J.*, **301**, 503.

Paczyński, B. (1986) "Gravitational microlensing and gamma-ray bursts", Princeton preprint.

Paczyński, B. (1986) "Gravitational micro-lensing by the galactic halo", *Astrophys. J.*, **304**, 1.

Refsdal, S., Grieger, B., and Kayser, R. (1987) *Gravitational Micro-Lensing as a Clue to Quasar Structure* to appear in the Proceedings of the 13th Texas Symposium on Relativistic Astrophysics.

Schneider, P. and Wagoner, R. V. (1986) "Amplification and polarization of supernovae by gravitational lensing", *Astrophys. J.*, **314**, 154.

Schneider, P. and Weiss, A. (1986) "A gravitational lens origin for AGN-variability?", *Astron. Astrophys.*, **171**, 49.

Subramanian, K., Chitre, S. M., and Narasimha, D. (1985) "Minilensing of multiply imaged quasars: Flux variations and vanishing images", *Astrophys. J.*, **289**, 37.

Young, P. (1981) "Q0957 + 561: Effects of random stars on the gravitational lens", *Astrophys. J.*, **244**, 756.

6. Compound Lenses

At least two of the multiply imaged quasars, $2016 + 112$ and $2237 + 031$, appear to have two lensing galaxies at substantially different redshifts. Furthermore, some models of galaxy formation predict the existence of large numbers of dark halo potential wells which, although too shallow to cause multiple imaging individually, may be able to do so by acting in concert. In this section, we consider the modifications that are brought about when there are two or more distinct lens planes.

Suppose we have a system of n lenses located on "screens", and we call the source, the $n + 1$ "screen" (see Fig. 24). Then, the time delay along a virtual path

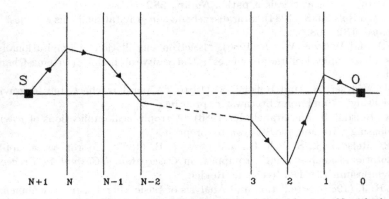

Fig. 24. An N screen lens. There are N deflecting planes labeled by 1 to N with the observer at point O and the source at point S. It is notationally convenient to label the observer as point number 0 and the source as point number $N + 1$.

parametrized by the intersection of the ray with the screens $(\vec{\theta}_i)$ is the sum of the contributions from each screen

$$t = \sum_{i=1}^{n}(1+z_i)\left[\frac{1}{2}\left(\vec{\theta}_{i+1}-\vec{\theta}_i\right)^2\frac{D_{0i}D_{0i+1}}{D_{ii+1}} - 2\phi_i^{(2)}(\vec{\theta}_i)\right] \tag{6.1}$$

where the D_{ij} are angular diameter distances in the usual notation, and $\phi_i^{(2)}$ is the two dimensional Newtonian potential of the ith screen. The position of the image for a given source is found by solving the $2n$ variational equations

$$0 = (1+z_i)(\vec{\theta}_{i+1}-\vec{\theta}_i)\frac{D_{0i}D_{0i+1}}{D_{ii+1}}$$

$$- (1+z_{i-1})(\vec{\theta}_i - \vec{\theta}_{i-1})\frac{D_{0i-1}D_{0i}}{D_{i-1i}} + D_{0i}(1+z_i)\vec{\alpha}_i \tag{6.2}$$

where $\vec{\alpha}_i = 2\nabla_{\vec{\theta}}\phi(\vec{\theta})/D_{0i}$. Solving the equation for the deflection between planes i and $i+1$,

$$(1+z_i)(\vec{\theta}_{i+1}-\vec{\theta}_i)\frac{D_{0i}D_{0i+1}}{D_{ii+1}} = -\sum_{j=1}^{i}(1+z_i)D_{0j}\vec{\alpha}_j. \tag{6.3}$$

Of course, we really want the position $\vec{\theta}_{i+1}$ in terms of the image position $\vec{\theta}_1$,

$$\vec{\theta}_{i+1} = \vec{\theta}_1 - \sum_{j=1}^{i}D_{0j}(1+z_i)\vec{\alpha}_j\sum_{k=j}^{i}\frac{D_{kk+1}}{D_{0k}D_{0k+1}(1+z_k)}$$

$$= \vec{\theta}_1 - \sum_{j=1}^{i}D_{j0}\vec{\alpha}_j\sum_{k=j}^{i}\frac{D_{kk+1}}{D_{k0}D_{0k+1}} \tag{6.4}$$

where we have used the property of angular diameter distances that

$$(1+z_j)D_{ij} = (1+z_i)D_{ji}.$$

The second summation over the expression of angular diameter distances can be simplified by examining the case in which only one $\vec{\alpha}_j$ to be non-zero. Then, Eq. (6.4) must simplify to the single lens plane equation (2.1) so that

$$\sum_{k=j}^{i}\frac{D_{kk+1}}{D_{k0}D_{0k+1}} = \frac{D_{ji+1}}{D_{j0}D_{0i+1}}. \tag{6.5}$$

This result can also be derived as a consequence of the following identity. For any well defined angular diameter distance, the four points 0, 1, 2, and 3 along a geodesic, obey the following relation:

$$D_{02}D_{13} = D_{01}D_{23} + D_{12}D_{03}.$$

This identity can be used recursively to prove Eq. (6.5). Hence,

$$\vec{\theta}_{i+1} = \vec{\theta}_1 - \sum_{j=1}^{i}\frac{D_{ji+1}}{D_{0i+1}}\vec{\alpha}_j. \tag{6.6}$$

The amplification tensor, defined by the differential change in the image postion for a change in the source position, is a non-linear product of the amplification tensors of each screen considered separately.

$$H_{\alpha\beta} = \frac{\partial \vec{\theta}_{n+1}}{\partial \vec{\theta}_1} = \delta_{\alpha\beta} - \sum_{j=1}^{n} \frac{D_{jn+1}}{D_{0n+1}} \frac{\partial \vec{\alpha}_j}{\partial \vec{\theta}_j} \frac{\partial \dot{\vec{\theta}}_j}{\partial \vec{\theta}_i} \tag{6.7}$$

which can be expanded recursively in terms of the single screen amplification tensors:

$$H_{\alpha\beta} = \delta_{\alpha\beta} - \sum_{k=1}^{n} \frac{D_{0k}D_{kn+1}}{D_{0n+1}} C_{\alpha\beta}^k + \sum_{k=1}^{n-1}\sum_{l=k+1}^{n} \frac{D_{0k}D_{kl}D_{ln+1}}{D_{0n+1}} \sum_{\gamma} C_{\alpha\gamma}^l C_{\gamma\beta}^k$$

$$- \sum_{k=1}^{n-2}\sum_{l=k+1}^{n-1}\sum_{m=l+1}^{n} \frac{D_{0k}D_{kl}D_{lm}D_{mn+1}}{D_{0n+1}} \sum_{\gamma}\sum_{\delta} C_{\alpha\gamma}^m C_{\gamma\delta}^l C_{\delta\beta}^k + \cdots \tag{6.8}$$

where the tensor

$$C_{\alpha\beta}^k = \frac{1}{D_{0k}} \frac{\partial \alpha_{(k)\alpha}}{\partial \theta_{(k)\beta}}. \tag{6.9}$$

While each of the individual 2×2 matrices $C_{\alpha\beta}^k$ is symmetric, their products are not. Hence, the magnification tensor is no longer symmetric so the images can be rotated relative to one another. There is no global scalar arrival time except in $2n$ dimensional space. In a few simple cases where all but one of the lenses are linear (characterized by a quadratic potential), or the multi-screen system has a high degree of symmetry, the lens system can be reduced to a single non-linear screen with an extra quadratic contribution to the potential.

In general, if we parametrize a virtual ray by the $2n$ coordinates $\vec{\theta}_n$, then we can form a general Hessian

$$H'(\vec{\theta}_n) = \frac{\partial^2 t}{\partial \theta_\alpha^i \partial \theta_\beta^j}. \tag{6.10}$$

Caustics are located where $|H'| = 0$.

References

Blandford, R. D. and Narayan, R. (1986), "Fermat's principle, caustics, and the classification of gravitational lens images", *Astrophys. J.*, **310**, 568.

Kovner, I. (1987), "The thick gravitational lens: A lens composed of many elements at different distances", *Astrophys. J.*, **316**, 52.

Schneider, P. and Borgeest, U. (1986), "Multiple scattering gravitational lens theory and general properties of gravitational light deflection", Max-Planck Institute #237, preprint.

7. The Observational Position

Gravitational lenses provide a theorists' heaven and an observers' hell.

— Anonymous

There are two aspects to be considered here. First, the geometry of the individual cases coupled with estimates of the *a posteriori* probabilities, and secondly, the statistical issues of lensing. We also append a summary of the space densities of sources and lenses.

7.1. *A Reprise of Existing Candidates*

7.1.1. *The double QSO: 0957 + 561*

This is the first gravitational lens system found, and the most heavily observed. We see two similar images of a quasar, A and B, with a separation of 6″ and a redshift of 1.41. The A image has optical (red) magnitude $m_r = 17.3$ and the B image is slightly fainter, with $m_r = 17.6$. The lensing galaxy has a redshift of 0.36, and is located about 1″ north of the B image. It is magnitude $m_r = 18.5$, fainter than the quasar images, and has ellipticity 0.13 at roughly 45 degrees to the line between the images. Other galaxies between A and B are ruled out to a limiting r magnitude of 25; however, the observed galaxy is the brightest of a large cluster centered near the lens. The cluster center is not well determined by present observations.

The quasar is a radio source that has been observed at both VLBI and VLA scales. In VLA maps, a faint radio jet is seen extending from the B image to the core of the lensing galaxy. The extended radio source provides an important constraint for lens models. Using VLBI, the two images can be shown to be similar core-jet sources.

This lens is crudely explained by the combination of a galaxy and a cluster which increases the separation from galactic scales (2″) to the observed 6″. The parities of the two observed images found by VLBI observations are those expected by the model. Time delays of approximately one year have been claimed, approximately as predicted, but await confirmation. The expected third image probably lies in the core of the galaxy, and is sufficiently deamplified (Sec. 3.3), or "captured" by a star (Sec. 5.2), to make it unobservable. The light that forms one of the observed images, B, passes through the galaxy and has a small probability of being microlensed.

7.1.2. *PG 1115 + 080*

In this case, we see four images of a redshift 1.72 quasar, two of which, A and A', are so close together that they were initially believed to be a single image. The A and A' images have a combined magnitude of 16.3 with a separation of 0.5″ and a position angle of 28 degrees measure clockwise from north. The magnitudes of the separate components are not yet well determined. The B image is located 1.8″ west and 0.1″ south of $A - A'$ and is approximately ten times fainter. The C image is located $1.4''$ west and $1.8''$ south of $A - A'$ and is approximately five times fainter. There is probably a magnitude 19.8 galaxy located $1.1''$ west and $0.1''$ south of $A - A'$. The four images are in the commonest image geometry for five images, caused by a central elliptical lens. The fifth image would again be deamplified or "captured" in the core of the lensing galaxy.

7.1.3. *2016 + 112*

In $2016 + 112$, three point sources of radio and emission lines are seen, as well as two galaxies and two extended emission line regions. The emission lines correspond to a quasar redshift of 3.3. The A image is the brightest, with magnitude 21. B is $3''$ west and $1.5''$ south of A with magnitude 21.5 in the emission line. The C' image is $2.0''$ west and $3.4''$ south of A with magnitude 22.9 in the emission line. The two regions of extended emission A' and B' are both magnitude 22.8, and they are located at $2.9''$ west, $2.0''$ north, and $5.8''$ west and $1.2''$ south of A. Two objects which cause the lensing are seen. C is a strong radio source located near the C' image, at $2.1''$ west, $3.2''$ south. It is not directly seen in the optical, but by using the optical flux ratios of A to C', it can be estimated to have $m_r \sim 26$. The $C - C'$ separation is only a few tenths of an arcsecond, but from the radio observations, C and C' are clearly distinct objects. D is probably a giant elliptical galaxy at $z \simeq 1$. It has an optical magnitude of approximately $M_b \simeq 22.6$

With three images, 2016 is the only lens which seems to satisfy the odd image theorem for non-singular lenses. Unfortunately, the image arrangement seems to preclude simple three image models because of the acute angle between the A and B images relative to the C image. The two patches of Ly α emission, A' and B', may or may not be lensed images of a single object. Of the two galaxies, C and D, only the redshift of D is known, and it is likely that the redshift of C will not be determined in the near future because it is so faint. Recently a model has been proposed (with both lenses at the same redshift) which has five images two of which are deamplified by the two galactic cores. The model requires rather contrived and extremely elliptical galactic potentials, and has difficulty preventing the generation of extra visible images. A good test of the model is whether it is consistent with VLBI observations. It seems very likely that a final model for this system will require lenses at different redshifts.

7.1.4. *2237 + 0305*

> ...the number of engagements that go on seems to me considerably above the proper average that statistics has laid down for our guidance.
>
> — Oscar Wilde in *The Importance of Being Earnest*

Located in a very large nearby spiral ($\sigma = 300\,\mathrm{km\,s^{-1}}$ at $z = 0.04$) are two images of a quasar approximately $1''$ apart. Because the galaxy is so close, it requires a large surface density $\Sigma_{\mathrm{crit}} = 3.5\,\mathrm{g\,cm^{-2}}$ to generate multiple images, and it is probably an example of a marginal lens. If the lensing is performed by a highly elliptical structure in the galaxy, such as a bar, only half of the critical surface density is required, and the topology is a lemniscate with the saddle lying in the bar. The third image will be hard to locate because the quasar is not a radio source, and there could easily be enough reddening to dispose of a third optical image. The mean magnification of the images may be of order 10–20. There may be a second lens in the system identified by an absorption line in the spectrum at $z \simeq 0.6$ (Huchra, private communication).

What is disturbing about $2237 + 0305$ is the probability of its occurrence. The approximate cross section for such a lens is $\sigma_L \sim 10$ square arc-seconds on the image plane, which implies a source plane cross section $\sigma_S \sim \sigma_L/M \sim 1$ square arc-second assuming a magnification of 10. In any square arc-second, we expect $\sim 10^{-6}$ quasars (Sec. 7.1), and if the lens was found in a survey of $\sim 10^4$ galaxies we estimate the *a posteriori* probability for the existence of such a lens to be 10^{-2}

7.1.5. *3C324*

$3C324$ is a distant ($z = 1.2$) and very luminous radiogalaxy. Recent observations of emission lines seem to indicate components coming from emission at a redshift $z = 0.85$. Filters centered on an emission line at $z = 1.2$ seem to show two components separated by $1.1''$ and possibly two additional faint components. Simple models based on this hypothesis suggest that $3C324$ is amplified by ~ 2 magnitudes. Further work is required to confirm this.

7.1.6. *1042 + 178*

Four radio images are seen roughly at the vertices of a diamond with a length of $\sim 2 - 3''$ across the diagonal. Each image is roughly three times fainter than the next brightest. In the optical observations a single strongly distorted image is seen. Clearly further observations are required.

7.1.7. *The dark matter lenses: 2345 + 007, 1635 + 267, and 0023 + 171*

There are three kinds of lies: lies, damn lies, and statistics.

— Benjamin Disraeli

In the $2345 + 007$ system, two objects are seen, separated by $7''$ with no sign of a lens. This could be interpreted as evidence for dark matter, but we would argue that it is not a lens. The spectra are not strikingly similar as compared to other lenses. If quasars and galaxies have approximately the same two point correlation function, then the probability for unrelated quasars to be separated by $6''$ is approximately 2×10^{-4}. The total number of QSOs found is approximately 6000 so we expect ~ 1 such correlated pair on the scale of 2345. We expect the probability of random association at different redshifts to be approximately the probability of correlated association at the same redshift, and hence we should find a few QSOs separated by $6''$ at different redshifts. This is in fact observed. A similar situation is found in 1635 + 267 where two objects are seen separated by 4″, again without any striking spectral similarity. This is probably a second example of a random spatial correlation.

The recently discovered binary quasar pair PKS1145 – 071 has two optical images separated by 4.2″ with an optical luminosity ratio of 2.5. The spectra of the images are remarkably similar, and the deduced redshifts are not grossly inconsistent with $\Delta z = 0$. These observations are remarkably similar to those available for 2345 + 007 and 1635 + 267, and hence we might claim it as another example of a gravitational lens. In this case, however, the brighter optical image is also a radio source, while the fainter one is not — the lower limit on the radio luminosity ratio is 500. In short, this cannot be a gravitational lens. Unfortunately, radio quasars consist of only 1– 10% of all quasars (depending on luminosity) so that clear cut distinctions between lenses and binaries will be more difficult in most cases. In an unbiased sample of binary quasars (i.e. one which will find all quasar pairs), for each optical-radio pair, such as PKS1145 + 071, we expect ∼10 optical-optical pairs such as 2345 + 007.

The third case, 0023 + 171 was found as part of a VLA survey searching for gravitational lenses. In the radio, two bright sources, A and B, which appear to be opposed radio jets are seen 5″ from a fainter unresolved source C. The peak flux ratios of $A : B : C$ are approximately 9 : 1.5 : 1. Optical counterparts are seen for the AB pair and C with R magnitudes of 23.1 and 21.9 respectively. The apparent redshift of the objects from the observed spectral lines is 0.94. There is no object in the field which may be acting as a lens, except a very faint extended emission region near the optical image of C. Because the object is so close, we should be able to see the lensing object unless the $M/L \gtrsim 1000$. While in an unbiased survey, radio-radio pairs should be extremely rare (∼1 for each 100 optical-optical pairs), this object (as well as PKS1145 + 071) was found by radio observations which will find no optical-optical pairs. Hence this object could also be a chance correlation, as the probability of finding a second radio source near a known radio source has a much higher probability than the *a priori* probability of finding two radio sources separated by 5″.

7.1.8. *1146 + 111*

> When you have eliminated the impossible, whatever remains, however improbable, must be the truth.
>
> — Sir Arthur Conan Doyle in *The Sign of Four*

This is the (in)famous pair of objects claimed to form a 157″ separation lens system. The pair is located in a region containing a large number of quasars. The enormous mass required to generate a splitting of 157″ should distort or amplify the other quasars, and cause an observable dip in the microwave background radiation. None of these effects is observed. There has been a great deal of controversy concerning the spectra of the two objects. This is largely irrelevent to the issue of lensing because even if they were two images of the same quasar the difference in transit times ($\sim$1000 years) is large enough to allow the quasar's spectrum to vary and the images to become dissimilar. If QSOs are clustered as galaxies the probability of random correlation at this scale is also 2×10^{-4} and the expected number of such correlations on the observed sample of quasars is $\sim$1. Again, there should be a few 1146-like objects observed which are not lenses.

7.2. Space Density of Sources — Quasars, Galaxies, and Radio Sources

The space density of sources is crucial to discussions of amplification biasing, and positive results are a consequence of the chosen source distribution. The generally accepted result is that quasars are uniformly distributed in redshift space from $z \sim 0.5$ to 2.5. The number of quasars per square degree brighter than visual magnitude B_s is approximated by

$$\log N(B_s) = \begin{cases} 0.5 + 0.9(B_s - 19) & 15 < B_s < 19 \\ 0.5 + 0.9(B_s - 19) - 0.14(B_s - 19)^2 & 19 < B_s < 22 . \end{cases} \quad (7.1)$$

This means that at 16^m there are 6.3×10^{-3} quasars per square degree, at 19^m there are 3.2, and at 22^m there are 100. There are approximately 4×10^6 quasars over the entire sky, and two to four thousand of them have been studied. Objects fainter than 22^m at $z \sim 2$ are generally considered to be Seyferts (the definition of quasar is generally taken to be $M_B < 23$ for $h = 0.5$).

7.3. Space Density of Lenses — Galaxies and Clusters

The distribution of galaxy lenses is derived from the local comoving density of galaxies. The fundamental assumption is that while the galactic luminosities may have undergone substantial evolution, the galactic potentials are largely unchanged from a redshift of approximately two to today. Nevertheless, this need not necessarily be true, if, for example, large galaxies develop from the merger of many smaller progenitors. Locally, the number density with luminosity L is well approximated by a Schechter function

$$\Phi(L)dL \simeq \frac{\phi_*}{L_*} \left(\frac{L}{L_*} \right)^{\alpha} e^{-L/L_*} dL \quad (7.2)$$

where

$$\alpha \simeq -1.3 \pm 0.3,$$
$$\phi_* \simeq (1.3 \pm 0.1) \times 10^{-2} h^3 \mathrm{Mpc}^{-3},$$
$$L_* \simeq 1.1 \times 10^{10} h^{-2} L_\odot.$$

The number of galaxies per unit comoving volume is

$$N = \int_{L_0}^{\infty} \Phi(L)dL \quad (7.3)$$

which is finite if $\alpha < -1$ with mean space density $\langle N \rangle = \phi_* \Gamma(\alpha + 1)$, and mean luminosity density $\langle L \rangle = \phi_* L_* \Gamma(\alpha + 2)$ where $\Gamma(x)$ is the gamma function. If we restrict ourselves to objects as luminous or more luminous than typical galaxies ($L_0 \simeq 0.2L^*$) or giant galaxies ($L_0 \simeq 3L^*$) the mean densities are somewhat lower. The comoving number density of galaxies at redshift z, corresponding to a local comoving density $N(0)$ is $N(z) = (1 + z)^3 N(0)$. The local density of rich clusters defined as having more than 50 bright galaxies within an Abell radius ($1.5\,h^{-2}\mathrm{Mpc}$) is $N(0) = (5\,h^{-1}\mathrm{Mpc})^{-3}$.

7.4. *Amplification Bias*

> He uses statistics as a drunken man uses lamp posts — for support rather than illumination.
>
> — Andrew Lang

As shown in Sec. 4.1, the cross section for a point source to have amplification greater than A is $\sigma \propto A^{-2}$. As discussed in Sec. 7.2, the bright quasar counts rise with flux decreasing S as $N(> S) \propto S^{-2.5}$ which suggests that multiply imaged systems are more likely to be highly amplified faint quasars than marginally amplified bright quasars. This is true only for sources brighter than 19^m. The sharp turnover in the quasar luminosity function above 19^m means that faint multiple images are more likely. As the brightest quasars are only about ten times brighter that 19^m quasars, amplification bias should enhance the incidence of multiple imaging by a factor of at most $\sim 10^{1/2} \sim 3$ at a given magnitude. It is not a large effect.

Ultimately the statistics of lenses may provide a simple test of the importance of amplification bias. General lenses are expected to be elliptical, and as discussed in Sec. 3.4, highly amplified images due to elliptical lenses are almost always five image systems. Hence, the ratio of five image systems to three image systems is an indication of whether or not the quasar counts do indeed turn over above 19^m as the Schmidt and Green luminosity functions indicate. A turnover in the luminosity function is indicated by a drop in the ratio of five to three image systems at lower observed luminosities. To the extent one can discuss statistics with four or five objects (a practice allowed only in astronomy), this is in fact what is observed. The two brightest lens systems are 1115 + 080 at 16^m and 2237 + 031 and 17^m. 1115 + 080 is the only known five image system (although 2016 + 112 may be one because of the added complication of the two lensing galaxies), and 2237 + 031 is a special case of a marginal lens.

7.5. *Surveys and Future Prospects*

Much of the promise of gravitational lensing has yet to be realized because of the paucity of lenses found to date. This will change, because there should be approximately one thousand lenses over the sky, and even finding fifty should begin to give adequate statistics to discuss galactic potentials and correlations. A series of surveys searching for multiple images on several scales should lead to better estimates of cross sections and optical depths which imply limits on Ω_{lens}. There are currently two surveys which are actively searching for gravitational lenses, and two new approaches which should soon be tested.

The VLA survey by Hewitt and collaborators is being used to search image splittings in the range of 0.1″–1″ and has found several candidates at the larger separation (2016, 0023, and 1042). These observations imply limits of $\Omega_{\mathrm{lens}} < 0.4$ at the smaller scale, and 0.7 at the larger. Similarly old VLBI data is being reanalyzed, looking for close pairs. An optical search is being conducted by Djorgovski and collaborators which produced the candidate 1635 + 267 and the binary pair PKS1154 + 071.

Automatic plate measuring (APM) offers the best means of finding galactic scale lenses. Current efforts involve examining Schmidt plates and seem to be very efficient down to 20^m. In a region of 130 square degrees, 5000 QSO candidates were found, of which 2500 were QSOs; 125 of 126 known QSOs in the field were found, along with 74 lens candidates of which 22 were considered to be good. A very significant result was that *none* of the candidates were found in the range 10″ to 2′ — that is to say, nothing like 1146 + 111 was seen. There is some hope of finding examples of lenses by kinking induced in the jets of steep spectrum quasars by foreground galaxies. There are now enough sources (about 50) and VLA maps to begin to examine the problem, but there is the difficulty of distinguishing the effects of lensing from intrinsic kinks in the jet. Finally, the Hubble Space Telescope (HST) may be extremely useful for finding faint lens galaxies and finding structure in the lenses but not in surveys to find new lenses. Such information is crucial for improved models of lenses which can be used to set constraints on galactic potentials.

References

0957 + 561

Booth, R. S., Browne, I. W. A., Walsh, D., and Wilkinson, P. N. (1979) "VLBI observations of the double QSO, 0957 + 561 A, B", *Nature*, **282**, 385.

Borgeest, U. and Refsdal, S. (1984) "The Hubble parameter: An upper limit from QSO 0957 + 561 A, B", *Astron. Astrophys.*, **141**, 318.

Dyer, C. C. and Roeder, R. C. (1980) "A range of time delays for the double quasar 0957 + 561 A, B", *Astrophys. J. Lett.*, **241**, L133.

Florentin-Nielsen, R. (1984) "Determination of difference in light travel time for QSO 0957 + 561 A, B", *Astron. Astrophys.*, **138**, L19.

Falco, E. E., Gorenstein, M. V., and Shapiro, I. I. (1985) "On model-dependent bounds on H_0 from gravitational images: Application to Q0957 + 561 A, B", *Astrophys. J. Lett.*, **289**, L1.

Falco, E. E., Gorenstein, M. V., and Shapiro, I. I. (1987) "Can we measure H_0 with VLBI observations of gravitational images?" in *IAU Symposium 129, The Impact of VLBI on Astrophysics and Geophysics*, eds. M. J. Reid and J. M. Moran (Dordrecht: Reidel), in press.

Gorenstein, M. V., Bonometti, R. J., Cohen, N. L., Falco, E. E., Shapiro, I. I., Bartel, N., Rogers, A. E. E., Marcaide, J. M., and Clark, T. A. (1987) "VLBI Observations of the 0957 + 561 Gravitational Lens System", in *IAU Symposium 129, The Impact of VLBI on Astrophysics and Geophysics*, eds. M. J. Reid and J. M. Moran (Dordrecht: Reidel), in press.

Gorenstein, M. V., Shapiro, I. I., Cohen, N. L., Corey, B. E., Falco, E. E., Marcaide, J. M., Rogers, A. E. E., Whitney, A. R., Porcas, R. W., Preston, R. A., and Rius, A. (1983) "Detection of a compact radio source near the center of a gravitational lens: Quasar image or galactic core", *Science*, **219**, 54.

Gorenstein, M. V., Shapiro, I. I., Rogers, A. E. E., Cohen, N. L., Corey, B. E., Porcas, R. W., Falco, E. E., Bonometti, R. J., Preston, R. A., Rius, A., and Whitney, A. R. (1984), "The milli-arcsecond images of Q0957 + 561", *Astrophys. J.*, **287**, 538.

Lebofsky, M. J., Rieke, G. H., Walsh, D., and Weymann, R. J. (1980) "The IR spectrum of the double QSO", *Nature*, **285**, 385.

Narasimha, D., Subramanian, K., and Chitre, S. M. (1984) "Gravitational lens model of the double QSO 0957 + 561 *A*, *B* incorporating VLBI features", *Mon. Not. R. Astron. Soc.*, **209**, 79.

Pooley, G. G., Browne, I. W. A., Daintree, E. J., Moore, P. K., Noble, R. G., and Walsh, D. (1979) "Radio studies of the double QSO, 0957 + 561 *A*, *B*", *Nature*, **280**, 461.

Porcas, R. W., Booth, R. S., Browne, I. W. A., Walsh, D., and Wilkinson, P. N. (1981) "VLBI structures of the images of the double QSO 0957 + 561", *Nature*, **289**, 758.

Shaver, P. A. and Robertson, J. G. (1983) "Common absorption systems in the spectra of the QSO pair Q0957 + 561 *A*, *B*", *Astrophys. J. Lett.*, **268**, L57.

Soifer, B. T., Neugebauer, G., Matthews, K., Becklin, E. E., Wynn-Williams, C. G., and Capps, R. (1980) "IR observations of the double quasar 0957 + 561 *A*, *B* and the interventing galaxy", *Nature*, **285**, 91.

Stockton, A. (1980) "The lens galaxy of the twin QSO 0957 + 561", *Astrophys. J. Lett.*, **242**, L141.

Walsh, D., Carswell, R. F., and Weymann, R. J. (1979) "0957 + 561 *A*, *B*: Twin quasistellar objects or gravitational lens?", *Nature*, **279**, 381.

Wills, B. J., and Wills, D. (1980) "Spectrophotometry of the double QSO, 0957 + 561", *Astrophys. J.*, **238**, 1.

Young, P., Gunn, J. E., Kristian, J., Oke, J. B., and Westphal, J. A. (1980) "The double quasar 0957 + 561 *A*, *B*: A gravitational lens image formed by a galaxy at $z = 0.39$", *Astrophys. J.*, **241**, 507.

Young, P., Gunn, J. E., Kristian, J., Oke, J. B., and Westphal, J. A. (1981) "Q0957 + 561: Detailed models of the gravitational lens effect", *Astrophys. J.* **244**, 736.

1115 + 080

Foy, R., Bonneau, D., and Blazit, A. (1985) "The multiple QSO PG1115 + 080: A fifth component?", *Astron. Astrophys.*, **149**, L13.

Christian, C. A., Crabtree, D., Waddell, P. (1987) "Detection of the lensing galaxy in PG1115 + 080", *Astrophys. J.*, **312**, 45.

Narasimha, D., Subramanian, K., and Chitre, S. M. (1982) "Spheroidal gravitational lenses and the triple quasar", *Mon. Not. R. Astron. Soc.*, **200**, 941.

Shaklan, S. B. and Hege, H. K. (1986) "Detection of the lensing galaxy in PG1115 + 080", *Astrophys. J.*, **303**, 605.

Weymann, R. J., Latham, D., Angel, J. R. P., Green, R. F., Leibert, J. W., Turnshek, D. A., Turnshek, D. E., amd Tyson, J. A. (1980) "The triple QSO PG1115 + 080: Another probable gravitational lens", *Nature*, **285**, 641.

Young, P., Deverill, R. S., Gunn, J. E., and Westphal, J. A. (1981) "The triple quasar Q1115 + 080 *A*, *B*, *C*: A quintuple gravitational lens image?", *Astrophys. J.*, **244**, 723.

2016 + 112

Lawrence, C. R., Schneider, D. P., Schmidt, M., Bennett, C. L., Hewitt, J. N., Burke, B. F., Turner, E. L., and Gunn, J. E. (1984) "Discovery of a new gravitational lens system", *Science*, **223**, 46.

Narasimha, D., Subramanian, K., and Chitre, S. M. (1984) "Gravitational lens models for the triple radio source MG 2016 + 112", *Astrophys. J.*, **283**, 512.

Narasimha, D., Subramanian, K., and Chitre, S. M. (1986) "Gravitational lens system 2016 + 112 revisited", *Astrophys. J.*, **315**, 434.

Schneider, D. P., Gunn, J. E., Turner, E. L., Lawrence, C. R., Schmidt, M., and Burke, B. F. (1987) "Spectroscopy of the extra-nuclear line emitting regions associated with the gravitational lens system 2016 + 112", Palomar Observatory, preprint.

Schneider, D. P., Gunn, J. E., Turner, E. L., Lawrence, C. R., Hewitt, J. N., Schmidt, M., and Burke, B. F. (1986) "The third image, the lens redshift, and new components of the gravitational lens 2016 + 112", *Astrophys. J.*, **91**, 991.

Schneider, D. P., Lawrence, C. R., Schmidt, M., Gunn, J. E., Turner, E. L., Burke, B. F., and Dhawan, V. (1985) "Deep optical and radio observations of the gravitational lens system 2016 + 112", *Astrophys. J.*, **294**, 66.

2237 + 031

Huchra, J., Gorenstein, M., Kent, S., Shapiro, I., Smith, G., Horine, E., and Perley, R. (1985) "2237 + 0305: A new and unusual gravitational lens", *Astrophys. J.*, **90**, 691.

Subramanian, K. and Chitre, S. M. (1984) "The Quasar 2237 +007 A, B: A case for the double gravitational lens", *Astrophys. J.*, **276**, 440.

Tyson, T. and Gorenstein, M. (1985) "Resolving the nearest gravitational lens", *Sky and Telescope*, **70**, 319.

1146 + 111

Bahcall, J. N., Bahcall, N. A., and Schneider, D. P. (1986) "Multiple quasars for multiple images", *Nature*, **323**, 515.

Blandford, R. D., Phinney, E. S., and Narayan, R. (1987) "1146 + 111 B, C: A giant gravitational lens?", *Astrophys. J.*, **313**, 23.

Crawford, C. S., Fabian, A. C., Rees, M. J. (1986) "Double clusters and gravitational lenses", *Nature*, **323**, 514.

Gott, J. R. (1986) "Is QSO1146 + 111 B, C due to lensing by a cosmic string?", *Nature*, **321**, 420.

Ostriker, J. P. (1986) "The effect of gravitational lenses on the microwave back-ground: The case of 1146 + 111 B, C", Princeton University, preprint.

Paczyński, B. (1986) "Is there a black hole in the Sky?", *Nature*, **321**, 419.

Phinney, E. S. and Blandford, R. D. (1986) "Q1146 + 111 A, B quasar pair: Illusion or delusion?", *Nature*, **321**, 569.

Shaver, P. A. and Christiani, S. (1986) "Test of the gravitational lensing hypothesis for the quasar pair 1146 +111 B, C", *Nature*, **321**, 585.

Turner, E. L., Schneider, D. P., Burke, B. F., Hewitt, J. N., Langston, G. I., Gunn, J. E., Lawrence, C. R., Schmidt, M. (1986) "An apparent gravitational lens with an image separation of 2.6 Arc minutes", *Nature*, **321**, 142.

Tyson, J. A. and Gullixson, C. A. (1986) "Colors of objects in the field of the double quasi-stellar object 1146 + 111 B, C", *Science*, **233**, 1183.

Other Lens Candidates

Chitre, S. M., Narasimha, D., Narlikar, J. V., and Subramanian, K. (1984) "3C273: A gravitationally lensed quasar?", *Astron. Astrophys.*, **139**, 289.

Djorgovski, S., Perley, R., Meylan, G., and McCarthy, P. (1987) "Discovery of a probable binary quasar", Center for Astrophysics, preprint.

Djorgovski, S. and Spinrad, H. (1984) "Discovery of a new gravitational lens", *Astrophys. J. Lett.*, **282**, L1.

Le Fèvre, O., Hammer, F., Nottale, L., and Mathez, G. (1987) "Is 3C324 the first gravitationally lensed giant galaxy?", *Nature*, **326**, 268.

Gott, J. R. and Gunn, J. E. (1974) "The double quasar 1548 + 115 A, B as a gravitational lens", *Astrophys. J. Lett.*, **190**, L105.

Hewitt, J. N., Turner, E. L., Lawrence, C. R., Schneider, D. P., Gunn, J. E., Bennett, C. L., Burke, B. F., Mahoney, J. H., Langston, G. I., Schmidt, M., Oke, J. B., and Hoessel, J. G. (1987) "The triple radio source 0023 +171: A candidate for a dark gravitational lens", preprint.

Kovner, I. and Milgrom, M. (1987) "Concerning the limit on the mean mass distribution of galaxies from their gravitational lens effect", *Astrophys. J. Lett.*, in press.

Paczyński, B. and Gorski, K. (1981) "Another possible case of a gravitational lens", *Astrophys. J. Lett.*, **248**, L101.

Shaver, P. A., Wampler, E. J., and Cristiani, S. (1987) "Quasar spectra and the gravitational lens hypothesis", *Nature*, **327**, 40.

Stocke, J. T., Schneider, P., Morris, S. L., Gioia, I. M., Maccacaro, T., and Schild, R. E. (1987) "X-ray selected AGNs near bright galaxies", *Astrophys. J. Lett.*, **315**, L11.

Surdej, J., Arp, H., Gosset, E., Kruszewski, A., Robertson, J. G., Shaver, P. A., Swings, J. P. (1986) "Further investigation of the pair of quasars Q0107 + 025 A and B", *Astron. Astrophys.*, **161**, 209.

Tyson, J. A., Valdes, F., Jarvis, J. F., Mills, A. P. (1984) "Galaxy mass distribution from gravitational light deflection", *Astrophys. J. Lett.*, **281**, L59.

Tyson, J. A., Seitzer, P., Weymann, R. J., Foltz, C. (1986) "Deep CCD images of 2345 + 007: Lensing by dark matter", *Astrophys. J.*, **91**, 1274.

Tyson, J. A. (1986) "Limits to galaxy mass distribution from galaxy-galaxy lensing", AT&T Bell Labs, preprint.

Weedman, D. W., Weymann, R. J., Green, R. F., and Heckman, T. M. (1982) "Discovery of a third gravitational lens", *Astrophys. J. Lett.*, **255**, L5.

Statistics and Amplification Bias

Avni, Y. (1981) "On gravitational lenses and the cosmological evolution of quasars", *Astrophys. J. Lett.*, **248**, L95.

Blandford, R. D. and Jarozyński, M. (1981) "Gravitational distortion of the images of distant radio sources in an inhomogeneous universe", *Astrophys. J.*, **246**, 1.

Canizares, C. R. (1981) "Gravitational focusing and the association of distant quasars with foreground galaxies", *Nature*, **291**, 620.

Kochanek, C. S. and Blandford, R. D. (1987) "Gravitational imaging by isolated elliptical potential wells II: Probability distributions", Caltech GRP-103, preprint.

Ostriker, J. P. and Vietri, M. (1986) "The statistics of gravitational lensing III: Astrophysical consequences of quasar lensing", *Astrophys. J.*, **300**, 68.

Ostriker, J. P. and Vietri, M. (1984) "Are some BL lacs artifacts of gravitational lensing?", *Nature*, **318**, 446.

Peacock, J. A. (1982) "Gravitational lenses and cosmological evolution", *Mon. Not. R. Astron. Soc.*, **199**, 987.

Schneider, P. (1984) "The Amplification Caused by Gravitational Bending of Light", in *Active Galactic Nuclei,* Dyson, J.E. (ed.), p. 351, Manchester: Manchester University Press.

Schneider, P. (1985) "Apparent number density enhancement of quasars near foreground galaxies due to gravitational lensing", *Astron. Astrophys.,* **143**, 413.

Schneider, P. (1986) "Statistical gravitational lensing and quasar-galaxy associations", *Astrophys. J. Lett.,* **300**, L31.

Schneider, P. (1986) "Statistical gravitational lensing: Influence of compact objects on the observed quasar luminosity function", Max-Planck Institute #218, preprint.

Schneider, P. (1986) "The Amplification-bias by gravitational lensing on observed number counts of quasars: The effects of finite source size", University of Colorado, preprint.

Turner, E. L. (1980) "The effect of undetected gravitational lenses on statistical measures of quasar evolution", *Astrophys. J. Lett.,* **242**, L135.

Turner, E. L., Ostriker, J. P., and Gott, J. R. (1984) "The statistics of gravitational lenses: The distributions of image angular separations and lens red-shifts", *Astrophys. J.,* **284**, 1.

Tyson, J. A. (1981) "Gravitational lensing and the relation between QSO and galaxy magnitude number counts", *Astrophys. J. Lett.,* **248**, L89.

Vietri, M. and Ostriker, J. P. (1983) "The statistics of gravitational lensing: Apparent changes in the luminosity function of distant sources due to passage of light through a single galaxy", *Astrophys. J.,* **267**, 488.

Vietri, M. (1985) "The statistics of gravitational lensing II: Apparent evolution in the quasars' luminosity function", *Astrophys. J.,* **293**, 343.

Galaxy Models, Quasar and Galaxy Distributions

Djorgovski, S. and Davis, M. (1986) "Fundamental properties of elliptical galaxies", *Astrophys. J.,* **333**, 59.

Djorgovski, S. (1987) "The Manifold of Elliptical Galaxies", to appear in *Structure and Dynamics of Elliptical Galaxies,* Proceedings of IAU Symposium 127, T. de Zeeuw (ed.), Dordrecht: D. Reidel.

Djorgovski, S. (1987) "Morphological Properties of Elliptical Galaxies", to appear in *Structure and Dynamics of Elliptical Galaxies,* Proceedings of IAU Symposium 127, T. de Zeeuw (ed.), Dordrecht: D. Reidel.

Kashlinsky, A. (1987) "Formation and Evolution of Clusters of Galaxies", to appear in *Continuum Radio Processes in Clusters of Galaxies,* O'Dea and Uson (eds.).

Kormendy, J. (1986) "Cores of early-type galaxies", Dominion Astrophysical Observatory, preprint.

Tremaine, S. (1987) "Summary", to appear in *Structure and Dynamics of Elliptical Galaxies,* Proceedings of IAU Symposium 127, T. de Zeeuw (ed.), Dordrecht: D. Reidel.

Surveys

Hewitt, J. N., Turner, E. L., Burke, B. F., Lawrence, C. R., Bennett, C. L., Langston, G. I., and Gunn, J. E. (1986) "A VLA Gravitational Lens Survey", to appear in *Observational Cosmology,* Proceedings of IAU Symposium 124, Dordrecht: D. Reidel.

Webster, R. L., Hewett, P. C., and Irwin, M. J. (1987) "An automated survey for gravitational lenses", University of Toronto, preprint.

8. Lenses as Probes of the Universe

It is much easier to be critical than to be correct.

— Benjamin Disraeli

So far, we have confined most of our discussion to the physics of lensing rather than its application to cosmology and the dark matter problem. As this is a school on "Dark Matter in the Universe", we should discuss the problem in this final lecture. At various points, we have alluded to applications of lensing to this problem, and here we will discuss some of these in greater detail. The real problem is that lensing is subject to as many ambiguities and statistical uncertainties as other approaches to the problem, so that there is no persuasive evidence for dark matter from any of these proposals.

8.1. *The Hubble Constant*

Studying modern literature concerning this topic, we find that especially (Anglo-) American authors (e.g. Young et al. 1981; Alcock and Anderson, 1985; Blandford, Narayan, and Nityananda, 1985) argue pessimistically about the reliability of the method outlined above. In my opinion this pessimism is unjustified.

— U. Borgeest, 1986, on determining H_0 or galactic masses

As we are all aware, values for H_0 range from 55 to 100 km s^{-1}Mpc^{-1} with formal errors of ± 5 km s^{-1}Mpc^{-1} and the question has remained this way for over a decade. The application of lensing to determining H_0 was suggested long before 0957 + 561 was discovered; today, the field is divided between the optimists, who believe that lensing will solve the problem, and the pessimists, who reply that the same has been said for all other approaches and that lensing is not going to remove the uncertainty.

Recall from Sec. 3.4 that the time delay between the two brightest images scales as

$$\Delta t_{12} \propto \frac{\theta_{12}^2}{h \bar{A}_{12}} \tag{8.1}$$

where the proportionality constant depends on the lens model and the cosmology, θ_{12} is the image separation, and $\bar{A}_{12}$ is the mean amplification of the two brightest images. The image separation can be very accurately measured for lenses which are radio sources and opposed optical images-allied optical images will have higher measurement errors because the separation scale is at best only marginally resolved. The mean amplification is not an observable, and must be determined from a lens model.

The two image lens systems, such as 0957 + 561, provide only three numbers to constrain a lens model and consequently the arrival time surface: the components of the separation vector and the ratio of the image amplifications. (Actually, 0957 + 561 provides several more constraints because of the radio jet.) The potential well itself is poorly determined by observations because the lens is not seen in all systems, and other local effects, such as the cluster in 0957 + 561, may be important. Furthermore, galactic and cluster potentials are poorly determined (otherwise there would not be a dark matter problem) and hence, given the position of the lens, there is still substantial freedom in modeling. A third image if it is observed, essentially gives the position of the lens and an estimate of the core size (from its deamplification, see Sec. 3.4). However, there is still a large degree of freedom in model selection, and hence a systematic error.

The simplest way to see this is to consider adding extra terms to the surface potential $\phi^{(2)}(\vec{\theta})$. Adding a constant changes neither the properties of the images nor the time delays associated with pairs of images. Adding a term which is linear in $\vec{\theta}$ is equivalent to adding a contribution to the time delays which is linear in the image separation. Linear terms correspond to adding a fixed bending angle at all images so that it has no effect on the image geometry or amplification. Adding a quadratic term, $a\theta^2$ associated with a uniform density sheet, introduces changes in the amplification of the images. However this change is equivalent to a change in the distance factor $D_{OL}D_{LS}\backslash D_{OS}$ and, consequently, to a change in the Hubble constant. Only if we assert, in the face of all evidence to the contrary, that the mass traces the light with a fixed mass to light ratio, can we hope to define the lens model well enough to give an accurate determination of H_0 (for a given cosmological model). The accuracy of this approach obviously improves as the number of constraints increases — ideally you want many images with known time delays, plus a well modeled lens galaxy (e.g. from knowing the rotation curve).

An additional problem concerns the choice of the angular diameter distance. For Friedmann models, the combination $D \equiv D_{OL}D_{LS}/D_{OS}$ is fairly insensitive to the choice of Ω_0. A typical example is that for $z_L = 0.5$ and $z_S = 3, D$ increases from 0.15 to 0.17 as Ω_0 decreases form 1 to 0 (see Fig. 25). However, if we model the universe as being Einstein–De Sitter on the large scale, but with all of the mass concentrated in large objects well removed from the beam which have a low probability of influencing the geodesic, the angular diameter distance is roughly the affine parameter λ defined in Sec. 2.3. In this case $D \simeq 0.12$. As the deduced value of $H_0 \propto D^{-1}$, we see that "emptying" the beam significantly alters the deduced value of the Hubble constant. In fact, the residual shear introduced by the matter near the beam can alter the travel time in an indeterminate way. Furthermore, if the universe is in any serious way inhomogeneous, then there is probably a fairly serious observational selection because multiple imaging will be favored along those lines of sight which happen to contain extra focusing matter. All in all, some fairly strong cosmological assumptions will have to be made to measure H_0.

8.2. *Galactic Masses*

As gravitational lensing couples directly to the gravitational potential of the lens it offers the hope of measuring the potential or the mass of the lens. In theory, lensing should give a value for the mass interior to the observed images. In practice, many of the difficulties in measuring the Hubble constant apply to measuring the mass; it is not so much the problem of finding a value as improving over other

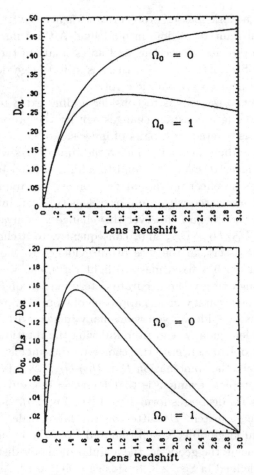

Fig. 25.   Comparison of angular diameter distances for different Ω_0. Top figure shows the angular diameter distance D_{OL} for Friedmann universes with $\Omega_0 = 0$ and 1. The figure at the bottom shows $D \equiv D_{OL}D_{Ls}/D_{OS}$ for the two cases. Note that D is much less sensitive to Ω_0 than D_{OL}.

techniques and understanding the systematic errors. (It is noteworthy that if the potential is modeled by a uniform density sheet and a circular galaxy model, it is possible to estimate the mass independent of the Hubble constant.) Detailed study of the gravitational lenses is difficult because, as Fig. 5 shows, most lenses will be at redshifts of 0.5 or greater where normal galaxies are too faint for good observations. Moreover, the real debate is over dark matter halos in galaxies which are expected to dominate the potential only on scales larger than the image splittings caused by an isolated galaxy. Hence lensing will tend to measure the luminous mass rather than the mass of a dark matter halo. In cases with larger splittings (such as $0957 + 561$) the effects of dark matter local to the lensing galaxy are difficult to separate from the external effects of the surrounding cluster.

Objects observed at large impact parameters will be distorted but not multiply imaged by a lens. By studying the effect of lensing on the apparent shapes of galaxies Tyson and collaborators have obtained a limit on the amount of matter in galaxies. This translates into an upper limit on rotation velocities of $v_{\text{circ}}(65\,\text{kpc}) < 190\,\text{km s}^{-1}$ which corresponds to $\Omega_{\text{gal}} < 0.03$, not inconsistent with conservative estimates for dark matter in galactic halos. However, this result is also sensitive to the cosmographic assumptions that are made.

8.3. *Lensing by Dark Matter*

Lensing by dark galaxies or unilluminated dark matter halos have been proposed as the explanation for the dark matter lenses described in Sec. 7.1.6. In particular, recent cosmological simulations of galaxy formation suggest the existence of dark halos of mass with central surface densities $\sim 0.6\sigma_{300}^2 r_{10}^{-1}$ gm cm^{-2}, which will require two or three aligned halos to generate multiple images. The major difficulty with this explanation is that the dark matter halos lack a deep core in which the third image of the dark matter lenses could be captured and de-amplified. Such lenses are likely to be marginal, and hence generate three equally bright images. Alternative explanations are that these dark lenses are due to "burnt out galaxies" in which an early elevated star formation rate leads to a non-luminous lensing galaxy at later times. A final suggestion is to make galaxies more compact at earlier times by having the potentials dominated by dark matter, which later decays and allows the galaxy to expand. This would lead to larger splittings and stronger deamplification of the odd image in the galactic core.

Compact objects either randomly distributed or in galactic halos are another possible type of dark matter. There have been a number of tests for dark matter in the mass range from $10^{-4}M_\odot$ to $10^6 M_\odot$. There are a few weak constraints from the observation that microlensing of field quasars is unimportant (because in most QSOs the bright emission lines have roughly the same equivalent width). This rules out $\sim M_\odot$ objects with $\Omega \sim 1$. Similarly, VLA surveys report finding no image splittings in the range of 0.1–1″ range which puts a limit of $\Omega < 0.4$. The current limit on compact objects in our galactic halo are that the optical depth must be $\tau < 10^{-6}$.

Giant black holes ($10^{15}M_\odot$) and cosmic strings both generate two images with large splittings. Both are detectable through a small dip in the microwave background radiation. This has been sought but not found in the case of 1146 + 111 for which both objects have been suggested. In fact, the automated plate measurement surveys limit the incidence of of multiple imaging with separations in the range from 10″ to 2′. This can be used to argue against large populations of objects more massive than galaxies which are capable of lensing. At present, all of the claimed lenses with no lensing galaxy are consistent with coincidence, so there is no reason to believe strings or other exotica are responsible.

Lensing caused by a cluster or supercluster would be interesting because it would provide some information on the mean density in the cluster or supercluster. However, remember that lensing really depends on a density contrast, and generally requires a fairly small core size to cause multiple imaging (see Eq. (1.3)). Even if clusters are capable of multiply imaging, the gains in cross section due to their large size are offset by their small number density. Recall that the probability of galactic lensing is of the same order of magnitude as the probability of a random association on the scale of arc seconds (if quasars cluster as galaxies). For clusters or superclusters, the probability of random association is probably orders of magnitude larger than the probability of lensing, and the time delays between images are too large for a correlation to confirm the existence of a lens. Lensing by clusters has been suggested as an explanation for the Lynds–Petrosian arcs near giant elliptical galaxies, but this is unlikely because of the small cross section and the extremely high lens symmetry required. The arcs cannot be due to multiple imaging because only a single arc is seen in each case, but one can imagine a scenario in which a single image is produced which has an enormous tangential amplification. Clusters and superclusters are generally too tenuous to image multiply and quasar pairs on this scale are probably nearly impossible to distinguish from coincidences.

The prospects of determinating Ω_0 directly from gravitational lensing are no brighter. Probably the best way to do this would be to locate the most probable lens redshift for a sample of high redshift lensed quasars. If we ignore any evolution in lens surface density, and assume a constant comoving density of lenses, the most probable redshift is at $z_L = 0.48$ if the source redshift $z_S = 3$. For $\Omega_0 = 0$ this changes to 0.56, which is indistinguishable (see Fig. 25) from the $\Omega_0 = 1$ average. Again, selection effects in the detectability of lenses threatens to introduce substantial bias.

References

Measuring Galaxy Masses and H_0

Alcock, C. and Anderson, N. (1984) "On the use of measured time delays in gravitational lenses to determine the Hubble constant", *Astrophys. J. Lett.*, **291**, L29.

Borgeest, U. (1983) "The difference in light travel time between gravitational lens images: II. Theoretical foundations", *Astron. Astrophys.*, **128**, 162.

Borgeest, U. (1986) "Determination of galaxy masses by the gravitational lens effect", *Astrophys. J.*, **309**, 467.

Cooke, J. H. and Kantowski, R. (1975) "Time delays for multiply imaged quasars", *Astrophys. J. Lett.*, **195**, L11.

Kayser, R. and Refsdal, S. (1983) "The difference in light travel time between gravitational lens images", *Astron. Astrophys.*, **128**, 156.

Kayser, R. (1986) "Gravitational lenses: Model fitting, time delay, and the Hubble parameter", *Astron. Astrophys.*, **157**, 204.

Kovner, I. (1987) "The marginal gravitational lens", *Astrophys. J.*, in press.

Kovner, I. (1987) "Extraction of cosmological information from multiimage gravitational lenses", *Astrophys. J. Lett.*, in press.

Refsdal, S. (1964b) "On the possibility of determining hubble's parameter and the masses of galaxies from the gravitational lens effect", *Mon. Not. R. Astron. Soc.*, **128**, 307.

Refsdal, S. (1966) "On the possibility of testing cosmological theories from the gravitational lens effect", *Mon. Not. R. Astron. Soc.*, **132**, 101.

Weinberg, S. (1976) "Apparent luminosities in a locally inhomogeneous universe", *Astrophys. J. Lett.*, **208**, L1.

Forms of Dark Matter

Canizares, C. R. (1982) "Manifestations of a cosmological density of compact objects in quasar light", *Astrophys. J.*, **263**, 508.

Dekel, A. and Piran, T. (1986) "Gravitational lenses and decay of dark matter", *Astrophys. J. Lett.*, **315**, L83.

Gott, J. R. and Park, M. (1986) "Limits on q_0 from gravitational lensing", Princeton University, preprint.

Gott, J. R. (1981) "Are heavy halos made of low mass stars? A gravitational lens test", *Astrophys. J.*, **243**, 140.

Kovner, I. (1987) "Subcritical gravitational lenses of large separation: Probing superclusters", *Nature*, **325**, 507.

Press, W. H. and Gunn, J. E. (1973) "Method for detecting a cosmological density of condensed objects", *Astrophys. J.*, **185**, 397.

Sanders, R. H., van Albada, T. S., and Oosterloo, T. A. (1984) "Gravitational imaging by superclusters", *Astrophys. J. Lett.*, **278**, L91.

Silk, J. (1986) "Have burnt-out galaxies and galaxy clusters been detected?", *Nature*, **324**, 231.

Subramanian, K. and Chitre, S. M. (1986) "The intensity ratio in different wavebands between images of gravitationally lensed quasars as a probe of dark matter", *Astrophys. J.*, **313**, 13.

Subramanian, K., Rees, M. J., and Chitre, S. M. (1987) "Gravitational lensing by dark galactic halos", *Mon. Not. R. Astron. Soc.*, **224**, 283.

9. Concluding Remarks

The discovery of at least four gravitational lenses over the past eight years has opened up a new subfield of extragalactic astronomy. However, we are probably attending the winter school under false pretenses, because gravitational lenses have not, as yet, illuminated the mystery of dark matter with which the school is concerned. Indeed, if our interpretation of the ten claimed examples of lensing is correct, it seems most likely that gravitational lenses will, at best, only be able to limit the presence of cosmic strings, intergalactic massive black holes, cold dark matter halos, etc. Nevertheless, in the case of 0957 + 561, gravitational lenses may provide confirmatory evidence that the centers of rich clusters do contain invisible matter with a velocity dispersion in excess of that associated with normal galaxies. In the other three good cases, the observations can be roughly explained in terms of isolated galaxies, although the details of the optics remain a matter of controversy.

From an observational standpoint, it is clear that more stringent spectral criteria must be satisfied than have been applied in the past before future claims of multiple imaging can be seriously entertained. In particular, the probability that similar quasars belong to the same group is not small. We would dearly like to have a better way of estimating this. The number of observed cases of multiple imaging is consistent with the lensing probability $\sim 10^{-3}$ obtained using standard galactic models. Many of the proposed uses and effects of lensing seem unpromising at present. There is no indication or expectation that lensing will seriously influence quasar counts, and neither is there yet any evidence for microlensing. We fear that ignorance of the details of the mass distributions in lenses and along the line of sight will thwart attempts to measure fundamental cosmological parameters (Ω_0 and H_0) with higher accuracy than other techniques.

Nevertheless, gravitational lenses remain full of fascination for the theorists. Various complementary approaches to the physics have been developed and applied. The challenge of accounting for the observations of 2016 + 112 and 2237 + 031 in detail remains. The development of an understanding of multiple screen propagation must be high on the theoretical agenda in view of evidence for this in both of these cases.

Gravitational lenses contribute one of the very few direct probes we have of distant matter, and in so far as understanding the evolution of structure in the universe is important, it is worthwhile to devote considerable amounts of observing time to discovering more examples of these phenomena and carefully reobserving the examples we have already found. As there are approximately 4×10^6 quasars over the sky, and the probability of lensing is $\sim 10^{-3}$, there remain ~ 4000 lenses to be found. Automatic plate surveys followed up by optical spectroscopy are a highly promising research technique, and if pursued, should furnish sufficient ($\gtrsim 30$) examples of lenses to pursue serious statistical and possibly cosmographic studies.

Acknowledgements

We would like to thank I. Kovner, S. Refsdal, and E. Falco for the extensive comments and corrections they have made to this manuscript.

Chapter 8

AN INTRODUCTION TO INFLATION

William H. Press[*]

Harvard-Smithsonian Center for Astrophysics
Cambridge, MA 02138, USA

David N. Spergel[†]

The Institute for Advanced Study
Princeton, N.J. 08540, USA

The previous lectures have considered the observational evidence for dark matter on various scales. This paper will consider that dark matter for which there is no observational evidence — the dark matter on very large scales needed to make $\Omega = 1$.

1. Review of Big Bang Cosmology

We first review the standard big-bang cosmology. We then explore the causal structure of the Robertson–Walker metric and find that different parts of our Universe are causally disconnected. In light of this, the observed isotropy of the Universe is very surprising and suggests either very special initial conditions or the need to modify the standard big-bang scenario. Inflation provides an attractive solution to this causality problem. Here, we introduce inflation and show how it provides a framework for producing perturbations that will form galaxies.

The equation for the time evolution of the expansion factor R in an isotropic homogeneous universe is

$$\left(\frac{\dot{R}}{R}\right)^2 = \frac{8\pi G\rho}{3} - \frac{K}{R^2},$$

(1)

where ρ is the energy density of the universe, and K is a constant. In a marginally unbound universe, $K = 0$. Otherwise the universe is clearly open or clearly closed, and $K = \pm 1$. We focus our attention on the "flat" $K = 0$ case. At early times, the $K \neq 0$ term was certainly negligible and, unless we live in a special epoch, it is still unimportant today. (Inflation will suppress this term, as we will see later in the lecture).

[*]Current address: Los Alamos National Lab., DIR/MSA-121, Los Alamos, NM 87545, USA.
[†]Current address: Princeton University Observatory, Princeton, NJ 08544, USA.

Throughout these lectures, we will use particle-physics units: $c = 1, k = 1$, and $\hbar = 1$. We will not, however, set $G = 1$.

The first law of thermodynamics describes the evolution of the energy density,

$$\frac{d\rho}{dR} = -\frac{3}{R}(p + \rho), \tag{2}$$

and we also need an equation of state for the pressure, $p(\rho, s)$. We will make the benign assumption that $p = p(\rho)$, but will remember that there will be glitches in $p(\rho)$ whenever one of the components of the universe becomes non-relativistic.

There are two equations of state usually encountered in cosmology. If the universe is matter-dominated or cold, then $p = 0$. Most people believe that our current Universe is in this state. Using Eqs. (1) and (2) yields,

$$\rho \propto R^{-3}, \qquad R \propto t^{\frac{2}{3}}. \tag{3}$$

If the universe is radiation dominated, then $p = \frac{1}{3}\rho$ and in this "hot" universe,

$$\rho \propto R^{-4}, \qquad R \propto t^{\frac{1}{2}}. \tag{4}$$

Later we will consider what happens when $p = -\rho$.

Now let us consider the causal structure of this expanding universe. We begin with the metric which describes the proper distance, s, between points in the universe,

$$ds^2 = -dt^2 + R^2(t)[dr^2 + r^2(d\theta^2 + \sin^2\theta d\varphi^2)]. \tag{6}$$

Observers on fixed coordinate lines ($\theta = \theta_0$, $r = r_0$, $\varphi = \varphi_0$) are unaccelerated. Since $ds^2 = -dt^2$, their time is proper time.

Two points in space-time are causally connected, if the proper distance between them is less than 0. This causal structure will be invariant under conformal transformations. That is, $ds^2 = 0$ can be multiplied by a non-singular function $f(t, r, \theta, \varphi)$ without doing any damage.

We will now use a series of conformal transformations in order to explore the causal structure. This procedure is called Carter–Penrose compactification. Rewriting the metric in terms of $d\eta = dt/R(t)$,

$$ds^2 = R^2(\eta)[-d\eta^2 + dr^2 + r^2d\Omega^2] \equiv R^2(\eta)d\bar{s}^2, \tag{7}$$

reveals that space-time in a Robertson–Walker $K = 0$ universe is conformally flat. This makes life more pleasant for field-theorists doing their calculation in a universe with this metric. Note that in a radiation dominated universe, $\eta \propto t^{\frac{1}{2}}$, while in a matter-dominated universe, $\eta \propto t^{\frac{2}{3}}$. In either case $R(t)$ is a power law and the interval $0 < t < \infty$ is mapped into $0 < \eta < \infty$.

Now switch to null coordinate, $v = \eta + r$ and $w = \eta - r$, and the flat piece of the metric can be rewritten as

$$d\bar{s}^2 = -dvdw + \frac{1}{4}(v - w)^2d\Omega^2. \tag{8}$$

The universe is now in an interval $-\infty < v < \infty$ and $-\infty < w < \infty$. This region can be compactified to fit on a sheet of paper by transforming to new coordinates:

$\tan p = v$, $\tan q = w$. The trigonometric identity,

$$\frac{(\tan p - \tan q)^2}{\sec^2 p \sec^2 q} = \frac{1}{4} \sin^2(p - q),\tag{9}$$

simplifies the metric:

$$d\bar{s}^2 = \sec^2 p \sec^2 q (-dp\,dq + \frac{1}{4}\sin^2(p-q)d\Omega^2).\tag{10}$$

The universe is now confined to the interval $-\frac{\pi}{2} < p < \frac{\pi}{2}$ and $-\frac{\pi}{2} < q < \frac{\pi}{2}$. It is revealing to undo the null coordinate, $t' = p + q$ and $r' = p - q$, and to write the metric in the final form:

$$d\bar{s}^2 = \sec^2\left(\frac{r+t}{2}\right)\sec^2\left(\frac{r-t}{2}\right)[-dt'^2 + dr' + \sin^2 r'\,d\Omega^2].\tag{11}$$

Figure 1a shows the universe confined to the interval $-\pi < t' + r' < \pi$ and $-\pi < t' - r' < \pi$. It is also bounded at $t = 0$ by a jagged line — the international signal for a singularity.

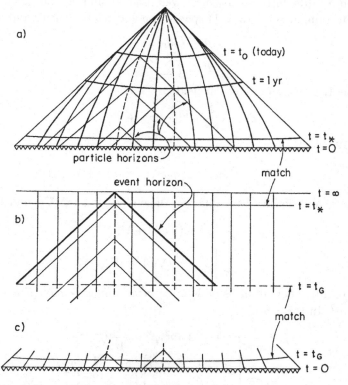

Fig. 1. Conformal diagrams showing causal structure (a) of a Friedmann cosmology, (b) of exponentially-expanding deSitter spacetime, (c) of an early Friedmann epoch. Guth's cosmology is obtained by "sewing together" parts of the three spacetimes, as indicated. See text for details.

The figure also shows the world-lines of galaxies in the universe. Note that at very late times $(t \to \infty)$, our light-cone will include the entire big bang. Our horizon is a particle horizon, since new bits of the universe keep crossing it. What is amazing is that these disconnected pieces of the universe were "pre-programmed" to have the same microwave background temperature. This requires either very special initial conditions or a little help from the particle physicists.

2. Inflation

Particle physics offers an attractive solution to the causality problem. During a phase transition, the universe undergoes a period of rapid expansion called "inflation" (Guth 1981, Albrecht and Steinhardt 1982, Linde 1982).

Connecting particle physics with cosmology requires that we derive an equation of state from a field theory. (See Piran (1986) for more detailed discussion.) The stress energy tensor can be calculated from a Lagrangian,

$$T_{\mu\nu} = \frac{-2}{\sqrt{-g}} \frac{\delta(\sqrt{-g}\mathcal{L})}{\delta g^{\mu\nu}} = -2 \frac{\delta \mathcal{L}}{\delta g^{\mu\nu}} + g_{\mu\nu}\mathcal{L}. \tag{12}$$

The relation $\delta(\ln(\det |g|)) = -g_{\mu\nu}\delta g^{\mu\nu}$ was used to simplify Eq. (12). Locally, we can switch to Minkowskian $(3 + 1)$ space-time and relate density and pressure:

$$\rho = T_{00},$$

$$p = \frac{1}{3}\sum T_{ii}. \tag{13}$$

Consider the Lagrangian of a scalar field φ:

$$\mathcal{L} = -\frac{1}{8\pi}[g^{\alpha\beta}\varphi_{,\alpha}\varphi_{,\beta} + V(\varphi)]. \tag{14}$$

Equation (12) yields

$$T_{\mu\nu} = \frac{1}{4\pi}\left[\varphi_{,\mu}\varphi_{,\nu} - \frac{1}{2}g_{\mu\nu}(\varphi_{,\mu}\varphi_{,\nu}g^{\mu\nu} + V(\varphi))\right]. \tag{15}$$

Thus, pressure and density have different dependencies on $\dot{\varphi}$, $\nabla\varphi$ and $V(\varphi)$:

$$\rho = \frac{1}{8\pi}\left[\dot{\varphi}^2 + (\nabla\varphi)^2 + V(\varphi)\right], \tag{16a}$$

$$p = \frac{1}{8\pi}\left[\dot{\varphi}^2 - \frac{1}{3}(\nabla\varphi)^2 - V(\varphi)\right]. \tag{16b}$$

Example: For a free field of mass m, $V(\varphi) = m^2\varphi^2$ and $\varphi \sim \exp[i(\vec{q}\cdot\vec{x} - Et)]$, where $E^2 = q^2 + m^2$. In this case,

$$\rho = \frac{1}{8\pi}[E^2 + q^2 + m^2]\varphi^2 = \frac{1}{4\pi}E^2\varphi^2, \tag{17a}$$

$$p = \frac{1}{8\pi}\left[E^2 - \frac{1}{3}q^2 - m^2\right]\varphi^2 = \frac{1}{12\pi}q^2\varphi^2. \tag{17b}$$

If the field is rapidly oscillating, $q \gg m$, and the universe is "hot": $p = \frac{1}{3}\rho$. On the other hand, if the field is slowly oscillating, $q^2 \ll m^2$, the universe is "cold": $p = 0$.

Now let us consider the case of adding a constant to $V(\varphi)$, that is, $V(\varphi) = V_0 + m^2\varphi^2$. This change will be unimportant when the universe is hot, $T \gg V_0$. However, later when $\varphi \to 0$, it modifies the equation of state:

$$\rho \simeq \frac{1}{8\pi}V_0 \equiv \frac{1}{8\pi}\Lambda_0, \tag{18a}$$

$$p \simeq -\frac{1}{8\pi}V_0 = -\frac{1}{8\pi}\Lambda_0 = -\rho, \tag{18b}$$

where Λ_0 is a cosmological constant. The stress energy tensor now has the form,

$$T_{\mu\nu} = \rho \begin{pmatrix} -1 & 0 & 0 & 0 \\ 0 & 1 & 0 & 0 \\ 0 & 0 & 1 & 0 \\ 0 & 0 & 0 & 1 \end{pmatrix}. \tag{19}$$

This Lorentz transforms like the identity matrix. It has no preferred state — it looks the same to all observers just like the vacuum state. Thus, this state is called a "False Vacuum" state. It has, however, a high gravitating density ρ.

Equations (1), (2), and (18) determine the cosmology of this False Vacuum state or De Sitter universe:

$$\frac{d\rho}{dR} = \frac{-3(p+\rho)}{R} = 0. \tag{20}$$

This implies the density of the universe does not change: $\rho = \rho_0$! The expansion equation,

$$\frac{\dot{R}}{R} = \sqrt{\frac{8\pi G\rho_0}{3}} \tag{21}$$

yields exponential expansion: $R \propto e^{Ht}$, where $H = (8\pi G\rho_0/3)^{\frac{1}{2}}$. In GUT theories, we might expect $\rho_0 \simeq aT_{\mathrm{GUT}}^4$. In this case,

$$H^{-1} = \left(\frac{T_{\mathrm{GUT}}}{M_{\mathrm{PLANCK}}}\right)^{-2} t_{\mathrm{PLANCK}} \simeq 10^{-35}\mathrm{s}. \tag{22}$$

For any particular theory, it is convenient to think of this as the unit time scale, the "tick": we will define 1 tick $\equiv 10^{-35}$s. The universe expands exponentially on this time scale.

Now let us consider causal structure of De Sitter space. Its metric,

$$ds^2 = -dt^2 + r_0^2 e^{2Ht}(dr^2 - r^2 d\Omega^2), \tag{23}$$

no longer has a power law relation between R and t. We will again change coordinates and study the causal structure. The transformation $\rho = r_0 r$, $\tau = (1-e^{-Ht})/H$ reveals that space–time is again conformally flat:

$$ds^2 = e^{2Ht}[-dt^2 + d\rho^2 + \rho^2 d\Omega^2]. \tag{24}$$

We have not compactified in space, but we have compactified the future time: $-\infty < \rho < \infty$, $-\infty < \tau < 1$. Figure 1b shows the world-lines of observers in De Sitter space.

These observers have event horizons, rather than particle horizons. Particles, which were in causal contact since the beginning of time, redshift away. Events in other galaxies that take place at late times lie forever outside our horizon.

We can learn more about the De Sitter metric through another change of variables:

$$r' = R_0 e^{Ht} r, \ t = t' + \frac{1}{2H} \log |H^2 r'^2 - 1|. \tag{25}$$

The transformed metric,

$$ds^2 = -(1 - H^2 r'^2) dt'^2 + \frac{dr'^2}{1 - H^2 r'^2} + r'^2 d\Omega, \tag{26}$$

reveals that De Sitter spacetime is stationary and looks something like an "inside-out Schwarzschild metric". Figure 1b shows the worldlines of homogenous isotropic observers in De Sitter spacetime. We will posit a pre-inflationary Robertson–Walker phase that generated these observers. This is shown in Figure 1c.

What sort of potential does inflation demand from the particle physics? At very high temperature, the universe should be in a radiation dominated De Sitter phase: $T \gg V(\phi)$. As the universe cools, the potential term becomes important and the universe enters the De Sitter phase with ϕ close to, but not exactly at the minimum of the potential. In the new inflationary scenario (Albrecht and Steinhardt 1982, Linde 1982), the Higgs field slowly rolls down the potential towards the true minimum. While it does so, the universe remains in a De Sitter phase. Since we want the universe to inflate by $\sim$ 60–70 e-folds, the universe must spend about 60–70 "ticks" in the rolling De Sitter phase. This requirement constrains $V''(\phi)$ to be small, but not unnaturally tiny. Eventually, ϕ reaches its true minimum value ϕ_0. Its oscillations around ϕ_0 are damped through its coupling to fermions. The Higgs field decays into these fermions and "reheats" the universe. Inflation does not solve the cosmological constant problem: It offers no explanation of why $V(\phi_0) = 0$.

What problems has inflation solved?

(1) Flatness problem: The K/R^2 term in the expansion equation,

$$\left(\frac{\dot{R}}{R} \right)^2 = \frac{8\pi G \rho_0}{3} - \frac{K}{R^2} \tag{27}$$

has been suppressed relative to the ρ_0 term by $R^2 \sim e^{120}$ during the De Sitter phase.

(2) Smoothness or causality problem: Figure 1 shows the worldlines of an isotropic homogenous observer as the universe evolves from an initial Robertson–Walker phase through a De Sitter phase and finally back to the present Robertson–Walker phase. Particles that were causally connected during the De Sitter phase cross the event horizon and drop out of causal contact. Later, the particles come back across the particle horizon in the Robertson–Walker phase. In the De Sitter phase, the universe had time to isotropize.

(3) Monopole problem: Inflation dilutes the number density of particles relative to ρ_0 by $R^3 > e^{200}$. It provides a mechanism for depopulating the universe of undesirable particles, topological defects and reducing any primordial anisotropy.

Inflation also provides a framework for generating gravitational perturbations. These perturbations can collapse to form galaxies. Cosmologists are interested in the spectrum of perturbations on scales between a few Kpc and the horizon scale ($2998h^{-1}$ Mpc). These physical scales correspond to quite a narrow range of timescales in the De Sitter phase, just a few "ticks".

It turns out that constant-amplitude perturbations which cross the outbound De Sitter event horizon — and whose evolution is thus "frozen" into purely kinematic behavior — reappear within the later Robertson–Walker particle horizon with a Zel'dovich (scale-free) perturbation spectrum. Since, *any* physical process that is slowly varying during the rolling phase (i.e. changes only over a timescale of 60 or 70 "ticks") is likely to be approximately stationary over the few "ticks" that created all scales of interest today, a Zel'dovich spectrum today is almost inevitable (Press 1981; Brandenberger, Kahn, and Press 1983; Brandenberger and Kahn 1984). The quantum fluctuations of "Hawking Radiation" are an example of such a stationary source of perturbations (Hawking 1982, Guth and Pi 1982, Starobinskii 1982; Bardeen, Steinhardt, and Turner 1983).

3. Additional Topics Not Covered Here

In the lectures on which this paper is based, several additional topics were reviewed: (i) details on the evolution of density perturbations outside and inside the horizon; (ii) the special case of cold dark matter, whose perturbations can start to grow even while the rest of the matter is locked in place by the radiation field; (iii) the relation between matter perturbations and the measured microwave anisotropy; and (iv) comparison with observations.

These topics have all been extensively discussed elsewhere, and little would be gained by another review here. For references, see Peebles (1982); Primack and Blumenthal (1984); Bardeen, Steinhardt, and Turner (1983); Brandenberger and Kahn (1983); Brandenberger (1985).

A small addendum to the standard literature on topic (ii) was discussed, namely the extension of the two-fluid formulas of Olive, Seckel, and Vishniac (1985) to the case where one fluid is the cooling, but not necessarily yet cold, dark matter, and the other fluid is the radiation (photon) background. (See also Bardeen 1980.) A very brief summary is as follows:

Choose units such that the expansion factor, now called a, equals unity when $\rho_m = \rho_\gamma$ (first moment of dark matter dominance), and take the unit of comoving wavenumber k to be the horizon size at that time.

The rate of growth of the density perturbation in dark matter, δ_m, depends upon the matter fluid velocity, v_m and viscous stress, Π_m:

$$\dot{\delta}_m = -(1+w)\frac{k}{a}v_m - 2w\frac{\dot{a}}{a}\Pi_m \qquad (28)$$

where $w = P_m/\rho_m$ is a function of temperature through the equation of state (see Service, 1986). When the matter is relativistic, $w = 1/3$, and when the matter is non-relativistic, $w \to 0$. The velocity perturbation grows in response to the matter overdensity:

$$\dot{v}_m = -\frac{\dot{a}}{a}v_m + \frac{k}{a}\Phi_A + \frac{k}{a}\frac{c_s^2}{1+w}\delta_m - \frac{2}{3}\frac{k}{a}\frac{w}{1+w}\Pi_m \qquad (29)$$

where $c_s^2 = dP_m/d\rho_m$ is the matter sound speed, also a function of temperature through the equation of state. The matter viscous stress depends only on the velocity perturbation. An adequate heuristic model of it is

$$\Pi_m = 4\sqrt{3}v_m\frac{(k/\dot{a})^2}{1+(k/\dot{a})^2}. \qquad (30)$$

The rate of growth of the photon density perturbation amplitude,

$$\dot{\delta}_\gamma = \frac{\dot{a}}{a}\delta_\gamma - \frac{4}{3}\frac{k}{a}v_\gamma - \frac{2}{3}\frac{\dot{a}}{a}\Pi_\gamma \qquad (31)$$

depends on the photon drift velocity, v_γ, and effective viscous stress Π_γ (which may actually be free-streaming or Landau damping). This photon viscous stress depends on the photon velocity in the same heuristic manner,

$$\Pi_\gamma = 4\sqrt{3}v_\gamma\frac{(k/\dot{a})^2}{1+(k/\dot{a})^2}. \qquad (32)$$

The photon velocity grows in response to the overdensity:

$$\dot{v}_\gamma = -\frac{\dot{a}}{a}v_\gamma + \frac{k}{a}\Phi_A + \frac{1}{4}\frac{k}{a}\delta_\gamma - \frac{1}{6}\frac{k}{a}\Pi_\gamma. \qquad (33)$$

The matter and radiation are coupled gravitationally by

$$\Phi_A = -\frac{3}{2}\left(\frac{a}{k}\right)^2\left[\frac{1}{2}\delta_m\frac{mw}{T}a^{-3} + \frac{1}{2}\delta_\gamma a^{-4}\right] - \left(\frac{a}{k}\right)^2\left[\frac{1}{2}\Pi_\gamma a^{-4} + \frac{1}{2}\Pi_m a^{-3}\left(\frac{mw}{T}\right)\right]. \qquad (34)$$

Here m is the rest mass of the dark matter, so the factor mw/T (which depends on the equation of state) is of order unity when the particle is non-relativistic.

Equations (28) through (34) are seven equations for the seven variables $\delta_m, \delta_\gamma, v_m, v_\gamma, \Pi_m, \Pi_\gamma$, and Φ_A. They can easily be integrated numerically in time to give the famous "Zel'dovich spectrum with a bend" (e.g. Blumenthal et al. 1984, Bond and Szalay 1983).

References

Albrecht, A. and Steinhardt, P. (1982) *Phys. Rev. Lett.*, **48**, 1220.

Bardeen, J. M. (1980) *Phys. Rev. D*, **22**, 1882.

Bardeen, J. M., Steinhardt, P., and Turner, M. (1983) *Phys. Rev. D*, **28**, 679.

Blumenthal, G., Faber, S. M., Primack, J. R., and Rees, M. J. (1984) *Nature*, **311**, 517.

Bond, J. R. and Szalay, A. S. (1983) *Astrophys. J.*, **174**, 443.

Brandenberger, R. (1985) *Rev. Mod. Phys.*, **57**, 1.

Brandenberger, R. and Kahn, R. (1983) *Phys. Rev. D*, **29**, 2172.

Brandenberger, R., Kahn, R., and Press, W. H. (1983) *Phys. Rev. D*, **28**, 1809.

Guth, A. (1981) *Phys. Rev. D*, **23**, 347.

Guth, A. and Pi, S.-Y. (1982) *Phys. Rev. Lett.* **49**, 1110.

Hawking, S. (1982) *Phys. Lett.* **B155**, 295.

Linde, A. (1982) *Phys. Lett.* **B108**, 389, and **B114**, 431.

Olive, K., Seckel, D. and Vishniac, E. (1985) *Astrophys. J.*, **292**, 1.

Peebles, P. J. E. (1982) *Astrophys. J. Lett.*, **263**, L1.

Piran, T. (1986) *Phys. Lett.*, **B181**, 238.

Press, W. H. (1981) in *Cosmology and Particles, Proceedings of the Moriond Astrophysics Meeting 1981*, J. Audouze et al., eds. (Dreux, France: Editions Frontieres).

Primack, J. R. and Blumenthal, G. R. (1984) in *Clusters and Groups of Galaxies*, ed. F. Mardirossian (Dordrecht: Reidel).

Service, A. T. (1986) *Astrophys. J.*, **307**, 60.

Starobinskii, A. (1982) *Phys. Lett.*, **B117**, 175.

Chapter 9

WIMPS IN THE SUN AND IN THE LAB

William H. Press*

Harvard-Smithsonian Center for Astrophysics,
Cambridge, MA 02138, USA

David N. Spergel[†]

Institute for Advanced Study,
Princeton, NJ 08540, USA

This lecture explores the hypothesis that weakly interacting massive particles (WIMPs) comprise the dark matter. Some kinds of proposed WIMP candidates can solve not only the missing mass problem, but also the solar neutrino problem. Future experiments may detect WIMPs.

1. WIMPS and the Solar Neutrino Problem

Over 60 years ago, Cecilia Payne, in her Harvard Ph.D. thesis, applied the newly discovered theories of atomic physics to the spectra of the Sun. She realized that the solar spectra suggested a disturbing discrepancy. While most of the Sun's spectral lines were metallic, the Sun seemed to be composed mostly of hydrogen and helium. This result ran counter to other evidence: meteors were mostly metallic and the Earth's crust was manifestly not composed of hydrogen. Further research has, of course, confirmed the work of Cecilia Payne: hydrogen and helium do compose 98% of the Sun and the heavier elements are merely tracers.

Today, something is again amiss. Our theoretical prejudices about inflation suggests that there ought to be sufficient matter to close the Universe, yet the baryon density inferred from primordial nucleosynthesis arguments accounts for only 10% of the closure density (Yang et al. 1984), if that. Within our own Galaxy, observations of the dynamics of stars and gas suggest that most of the mass is in non-luminous forms (Hegyi and Olive 1987). Even within the disc of our Galaxy, 50% of the matter is in some unknown forms (Bahcall 1984, Bahcall 1987). Perhaps matter as we know it — protons, neutrons, and electrons — is only a test tracer of the underlying distribution of the mass of the Universe.

Closer to home, the Sun is not behaving comprehensibly. From the Sun's luminosity, mass, and surface composition, and our understanding of stellar interiors,

---

*Current address: Los Alamos National Lab., DIR/MSA-121, Los Alamos, NM 87545, USA.
[†]Current address: Princeton University Observatory, Princeton, NJ 08544, USA.

we can infer the nuclear reaction rates in the solar core. These reactions should produce a high-energy neutrino flux, yet Ray Davis' chlorine detector in the Homestake Mine observes only one-third of the predicted flux (Bahcall et al., 1982).

One of the joys of astrophysics is that stars and galaxies are complicated. There are many possible sources for the discrepancy between theories of the solar interior and the neutrino observations: nuclear physics, hydrodynamics, or perhaps, particle physics. The candidates for the non-luminous matter range from 10^6 solar mass black holes through very low mass stars down to 10^{-6} eV axions. However, like the apocryphal drunk searching for his keys under the streetlight, we will attempt in this paper to find the non-luminous matter in the Sun. This review thus explores the hypothesis that unseen matter is the source of the "solar neutrino problem." It hypothesizes a massive weakly interacting particle,[a] and explores the astrophysical consequences of that particle. It also describes experiments that could detect these particles if they are indeed the "missing mass."

As we have learned in previous lectures, "missing mass" is, of course, a misnomer. What is missing is not the mass, but the light. Since from a dynamical point of view our Galaxy is a mature system, observations of the kinematics of tracer populations reveal the underlying mass distribution. Radio observations of atomic hydrogen in the discs of spiral galaxies provide evidence that the density distribution of galaxies falls off as r^{-2}. The atomic hydrogen flares out near the end of the disc, confirming that beyond 10 Kpc the dominant component is spheroidal. On the other hand, the light in the Galaxy is concentrated in the disc. The objects that compose the halo must be almost entirely non-luminous, and non-dissipative. WIMPs satisfy these conditions. (See Primack (1986) for a review of particle physics dark matter candidates.)

Now let us return from the distribution of mass in the Galaxy to the seemingly unrelated problem of the generation of energy in the Sun. The Sun generates it luminosity through the fusion of hydrogen into helium. There are three possible channels through which four protons can be converted into the two proton and two nucleons of a helium nucleus. One of these channels, PPIII, involves the decay of B^8 into Be^8, a positron, and an energetic neutrino. The neutrino flux from this reaction in the Sun can be detected on earth through the inverse beta decay of chlorine into argon. R. Davis and his collaborators, using a tank of 610 tons of carbon tetrachloride in the Homestake mine, have measured this flux and found that the observed flux is only one third of that predicted from standard solar models (Bahcall et al., 1982).

Because all of the reaction rates in hydrogen fusion are limited by the need to overcome the Coulomb repulsion of the nucleus, the relative rates of the three PP channels in the Sun are very temperature dependent. Any mechanism which cools the core of the Sun would reduce the rate of the reaction and hence the

---

[a]We will refer to any weakly interacting massive particle as a WIMP and to those particles that have the properties needed to solve the solar neutrino problem as "cosmions."

observed solar neutrino flux. We realized that weakly interacting particles could be extremely efficient at energy transport (Spergel and Press 1985). After publication, we learned that John Faulkner and Ron Gilliland had considered this possibility several years earlier, but, lacking at the time a plausible mechanism for getting the particles into the Sun, never published. Most of their conclusions were, however, summarized in Stiegman et al. (1978) in a subsection forgotten even by Faulkner. Their paper finally appeared seven years later (Faulkner and Gilliland 1985).

Particles with cross-sections of order 10^{-36} cm^2 are ideal for transporting energy in the Sun. In the conductive (large cross-section) regime, energy transport scales as the mean free path. As the cross-section decreases, the cosmion travels through a larger temperature gradient between collisions and is thus more effective at transporting energy. In the small cross-section regime, collisions are so rare that the energy transport scales as the collision rate. The cross-over between these two regimes occurs at the optimal cross-section for energy transport: when $\sigma \approx 10^{-36}$ cm^2 and the cosmion's mean free path is about its orbital radius. The cosmion can deposit a large fraction of its kinetic energy at aphelion and can increase its kinetic energy at perihelion.

The timescale for the cosmion to transfer energy from the center of the Sun to a cosmion scale height, the free fall time (≈ 100 s), is much shorter than the timescale for photons to diffuse the same distance, the Kelvin-Helmholtz time ($\approx 10^6$ years). A number density of cosmions of only 10^{11} is sufficient to significantly alter energy transport in the solar core and lower the predicted SNU flux to the observed value.

The net effect of the cosmion on the temperature distribution in the Sun is to cool the central core of the Sun while heating the region near the aphelion of the typical orbit (Spergel and Press 1985, Nauenberg 1986 and Gould 1987a). The scale height of the cosmion distribution can be estimated by equating the cosmion's thermal energy with its potential energy,

$$r_x = 0.13 \left(\frac{m_p}{m_x} \right)^{1/2} R_\odot.$$

Most of the B^8 neutrinos are produced in the inner $0.05 R_\odot$, while most of the Sun's luminosity is produced in the inner $0.2 R_\odot$. Thus a cosmion with mass between 2 and 10 GeV will reduce the B^8 neutrino production rate without reducing the solar luminosity or affecting the production rate of pp neutrinos. Hence the predicted count rate from a solar model *cum* cosmions for the pp-neutrino sensitive Ga^{71} experiment does not differ significantly from a standard model. Cosmions more massive than 10 GeV will be too centrally concentrated to affect the thermal structure in most of the B^8 neutrino producing region (Gilliland, Faulkner, Press and Spergel 1986).

The Sun will capture weakly interacting particles from the galactic halo. (This is the point that originally eluded Gilliland and Faulkner, but which we were fortunate enough to recognize.) The escape velocity from the Sun's surface is 617 km/s, while the escape velocity from the core is over 1000 km/s. A halo cosmion with

typical velocity 30 km/s will fall into the Sun where it can be captured through
a single collision as long as its mass is less than $\sim$50 proton masses. The Solar
capture rate is approximately the geometrical rate $\pi R_\odot^2 nv$, where v is the typical
halo cosmion velocity and n the number density, times the gravitational focussing
factor $(GM_\odot)/(R_\odot v^2)$. Press and Spergel (1985) discuss the various effects and find
that the capture rate is sufficient for the Sun to accumulate a significant number
of cosmions in the solar lifetime. If we multiply the capture rate by the lifetime, we
find that we can achieve a significant concentration of cosmions relative to baryons,

$$\frac{n_x}{n_b} \simeq 3 \times 10^{-10} \left(\frac{\rho_x}{1 M_\odot/pc^3} \right) \left(\frac{v_{\text{esc}}}{\bar{v}} \right) \left(\frac{\sigma}{\sigma_{\text{crit}}} \right) \left(\frac{m_p}{m_x} \right).$$

Recall that a concentration of 10^{-11} of cosmions with cross-section of 4×10^{-36} cm^2
can resolve the solar neutrino problem. If the cosmions compose the halo ($\rho_{\text{HALO}} \approx$
$10^{-2} M_\odot/pc^3, v_{\text{HALO}} \approx 300$ km/s), then their cross-section must be within a factor
of 2 of σ_{crit}. If cosmions compose the disc ($\rho_{\text{DISC}} \approx 10^{-1} M_\odot/pc^3, v_{\text{DISC}} \approx 50$ km/s),
then they might resolve the solar neutrino problem, if their baryon scattering cross-
section is between 10^{-37} and 10^{-34} cm^2.

Halo particles will also be captured by the Earth and the other planets (Freese
1986, Gould 1987b and Bouquet and Salati 1987). For a particle to be capturable,
its velocity at infinity must be less than the escape velocity from the planet's core.
Thus for the *capturable* particles, gravitational focusing is always important, even if
it is not important for the average particle in the distribution. Note that for stars,
for whom most of the flux is capturable, $F \propto v_{\text{esc}}^2 R^2 \propto MR$, while for planets,
which can capture only a fraction of the flux, $F \propto v_{\text{esc}}^4 R^2 \propto M^2$.

Cosmions do not significantly alter main sequence stellar evolution. Since they
are centrally concentrated within the luminosity producing region, they do not alter
the mass of available fuel (hydrogen). Many other proposed solutions to the solar
neutrino problem that reduce the thermal gradient in the inner core also reduce the
gradient throughout the luminosity producing region, thus increasing its volume.
With more available hydrogen, the main sequence lifetime of a star is longer. Such
solutions aggravate the discrepancy between the Hubble time and the inferred age of
globular clusters. Cosmions intrinsically have little effect in the evolution of massive
stars: Since their number density in the star grows linearly with time, few cosmions
can accumulate in a massive star's short lifetime.

Renzini (1986) argues that cosmions can significantly alter the evolution of horizontal branch stars. Faulkner and Spergel (1987); however, point out that Renzini
severely overestimated the cosmion energy transport. Cosmions are only effective
at energy transport when their mean free path is comparable to an orbital length.

Cosmions can significantly alter the evolution of old stars with strong temperature gradients and strongly temperature-dependent nuclear reactions. Pre-white
dwarfs cool through neutrino emission from their inner core (Lamb and van Horn
1975). This produces a strong temperature inversion (the core temperature of
3.3×10^8 K is much lower than the maximum temperature of 5.2×10^8 K at a

radius of 4.4×10^8 cm). The scale height of cosmions in thermal equilibrium with the carbon nuclei in the core is $5.3 \times 10^8 \sqrt{m_p/m_x}$ cm. Pre-white dwarfs might seem like interesting candidate objects; however, preliminary estimates indicate that an insufficient number density of cosmions will be accumulated to alter their evolution. Cosmion energy transport is about 10^{-2} times less effective than conductive energy transport.

Since cosmions alter the solar thermal structure, they affect the seismology of the Sun. Solar seismology, which measures the sound speed as a function of radius, might detect the variations in density and temperature induced by cosmion energy transport. Dappen et al. (1986) and Faulkner et al. (1986) suggest that cosmions can eliminate the discrepancy between the observed p-mode (pressure dominated) spectrum and the standard solar model. Bahcall and Ulrich (1987) argue that this discrepancy is not significant. Cosmions would have more dramatic effects on the still unobserved g-mode (gravity dominated) spectrum, which is more sensitive to the core conditions.

The good news is thus that a halo population of cosmions of the correct cross-section ($\approx 10^{-36}$ cm^2) and mass (5–10 GeV) will be captured by the Sun in sufficient number to resolve the SNU problem, and yet will not alter other aspects of stellar evolution. We now must turn to the bad news: Once captured by the Sun, cosmions can be lost either through annihilation or through evaporation.

Most of the cosmions in the Sun are tightly bound: their typical velocities, $\sqrt{3kT/2m_x} \approx 300$ km/s, is much less than the escape velocity from the core $v_{\text{esc}}^2 = 1400$ km/s, so scatterings that produce $v \geq v_{\text{esc}}$ are rare. In the conclusion of Spergel and Press (1985), the evaporation rate is estimated as the fraction of cosmion distribution with energy sufficient to escape divided by the time to repopulate the tail. This estimate suggests that evaporation is negligible for cosmions with $m > 4m_p$.

The cosmion distribution function may differ significantly from Maxwellian in its high energy tail. Nauenberg (1986), Greist and Seckel (1987) and Gould (1987a) have considered this effect and conclude that this lowers the evaporation limit to $3.5\,m_p$.

Annihilation can also reduce the number of cosmions in the core of the Sun. If the cosmion is a Majorana particle, it is its own anti-particle and will self-annihilate. If the cosmion is a Dirac particle and the Sun contains both it and its anti-particle in equal numbers, annihilation will also reduce its solar abundance. The cosmion annihilation timescale in the Sun can be estimated,

$$t_{\text{ann}} = (n_x \sigma_{\text{ann}} v)^{-1} = \left(\frac{n_p}{n_x}\right) \left(\frac{\sigma_{\text{ann}}}{\sigma_{bx}}\right) t_{\text{coll}},$$

where σ_{ann} is the cosmion annihilation cross-section and σ_{bx} is the cosmion-baryon scattering cross-section. If the cosmion is to resolve the Solar neutrino problem

$$t_{\text{coll}} \approx t_{\text{dynamical}} \approx 100\,\text{seconds}$$

and $n_p/n_x \approx 10^{11}$. Most of the attractive particle physics cosmion candidates (photinos, scalar neutrinos, massive and Dirac neutrinos) have scattering cross-sections less than or on the order of their annihilation cross-sections; this implies $t_{\text{ann}} < 10^{13}$ seconds, much shorter than the age of the Sun (Krauss, Freese, Spergel and Press 1986). Cosmions are more centrally concentrated than baryons; this enhances their annihilation rate and exacerbates the problem.

Even if annihilation reduces the solar abundance of these particles below the threshold for affecting the thermal structure of the Sun, these particles can still have observable effects in the Solar System. Silk, Olive and Srednicki (1985) suggest that the annihilation of 5 GeV photinos captured by the Sun from a halo population would produce a detectable gamma ray flux. Their calculation; however, assumes that the photino distribution function is well approximated by a Maxwellian even in the high energy tail. Freese (1986) and Krauss, Wilczek and Srednicki (1986) argue that if the halo is composed of massive or scalar neutrinos with masses greater than 12 GeV, a sufficient number density would be captured and retained by the Earth for annihilation in the Earth's core to produce a detectable flux of neutrinos in the IMB detector. Evaporation sets the 12 GeV cut-off and is sensitive to the Earth's core temperature. Gaisser et al. (1986) attempt a detailed treatment of the capture and annihilation of SUSY particles in the Sun. Gould (1987b) reconsiders WIMP capture and shows that previous authors have underestimated capture rate in the Earth. Silk and Srednicki (1984) and Stecker, Rudaz and Walsh (1985) suggest that WIMPs may also be detectable through anti-protons produced by their annihilations in the halo.

Requiring that annihilation does not greatly reduce the cosmion concentration in the Sun places severe constraints on cosmion particle physics. Either the cosmion annihilation cross-section in the Sun is much lower than its scattering cross-section or there is a net asymmetry between cosmions and anti-cosmions that reduces the number of anti-cosmions in the Sun. Either hypothesis constrains the physics of the cosmion sector and eliminates several particle candidates.

Gelmini et al. (1987) suggests that the "annihilation problem" can be resolved by imposing a net asymmetry between the cosmions and anti-cosmions. Either a net cosmological overabundance of cosmions or the cosmion-baryon scattering cross section exceeding the anti-cosmion–baryon scattering cross section will result in more cosmions in the Sun than anti-cosmions. Thus, while annihilation will eliminate the anti-cosmions, some of the cosmions will remain.

A slight asymmetry between the production rate of baryons and anti-baryons is often said to be responsible for the existence of large numbers of baryons in our Universe. (Hence, we exist to call it our Universe.) If the same mechanism functions for both baryons and cosmions, so that every net baryon is paired by a conservation law to a net cosmion, then their relative densities depend only on their mass:

$$\frac{\rho_{\text{cosmion}}}{\rho_{\text{baryon}}} = \frac{m_{\text{cosmion}}}{m_{\text{baryon}}}.$$

Primordial nucleosynthesis arguments yield $\rho_{\text{baryon}}/\rho_{\text{closure}} \approx 0.05$–$0.2$ (Yang et al., 1984). Thus, if the cosmions close the Universe, their mass lies between 4 and 20 GeV, the required range to effectively transport energy in the Sun. Raby and West (1987a) suggest that this asymmetry could produce a Universe dominated by fourth generation neutrinos with large magnetic moment that they have named "magninos". These particles have the needed properties to solve the solar neutrino problem and may be detectable in accelerator experiments (Raby and West 1987b).

If there is no asymmetry between cosmion and anti-cosmion number densities, the relic density is determined by the cosmion annihilation cross section. When the Universe becomes optically thin to cosmions (called cosmion "freezeout"), their annihilation ceases and their density per comoving volume reaches a constant. The cosmion density depends strongly on annihilation rate. For example, 5 GeV particles will close the Universe if their annihilation rate at decoupling is $\approx 8 \times 10^{-38} h_{100}^{-2} \text{ cm}^2$.

If, on the other hand, annihilation is not to deplete the cosmion number density in the Sun, $\sigma\beta(1 \text{ KeV}) \leq 10^{-42} \text{ cm}^2$. Thus we require a mechanism that will not only reduce the cosmion annihilation cross section in the Sun to less than 10^{-6} of the cosmion-baryon scattering cross section, but will also suppress low temperature annihilation by 10^{-5} relative to the annihilation rate at decoupling ($T \approx$ GeV). A lower cross section at the decoupling temperature would result in too large an abundance of cosmions relative to baryons.

S-wave suppression provides a mechanism for depressing the annihilation rate at low energies. Majorana particles will annihilate only with particles with opposite helicity states. The "s-wave" term in the annihilation cross section is thus no longer proportional to the square of the mass of the cosmion but rather the square of the mass of its decay products. The remaining "p-wave" term is velocity-dependent, $\sigma\beta$ (p-wave only) $\propto v^2 \propto T$ and thus low temperature annihilation is suppressed. Gelmini et al. (1987) describe a theory that exploits this suppression mechanism.

An intriguing possibility is that the cosmion might be a supersymmetric particle. Supersymmetry is the only symmetry of the extended Poincare group that has no observed experimental implications. It implies that every fermion has a supersymmetric bosonic partner and every boson has a supersymmetric fermionic partner. Supersymmetry resolves many theoretical particle physics problems. In many supersymmetric theories, there is a new conserved quantum number called R parity. This implies the existence of a new stable particle, since the lowest mass supersymmetric particle cannot decay into a lower mass particle with a different R parity. This particle, however, can still annihilate with an anti-particle of opposite R parity. All known particles have one R parity, while their (yet unobserved) supersymmetric partners have another R parity. For photinos (the supersymmetric partner of the photon), this annihilation occurs through the exchange of a squark (the partner of the quark) or the selectron (the partner of the electron). In many supersymmetric theories, the selectron and squark masses are smaller than but of the same order as the W mass. This implies that the photino will close the Universe if their mass

is approximately 5 GeV. Krauss et al. (1986) found that despite the suppression
of s-wave photino annihilation into up and down quarks, photino annihilation into
more massive quarks and leptons would reduce the concentration in the Sun below
that needed to affect the high-energy neutrino flux. Sneutrino annihilation would be
more rapid than photino annihilation; hence, sneutrinos are not attractive cosmion
candidates (Greist and Seckel 1987).

In conclusion, the cosmion hypothesis is beautiful astrophysics, but contrived
particle physics. If cosmions exist, they can solve several long standing problems
in astrophysics. Several particle physics models have been proposed for cosmions;
however, none of these models were motivated by some profound symmetry such as
supersymmetry. What makes the cosmion hypothesis exciting is that it is falsifiable.
The next section describes how these particles could be detected in future laboratory
experiments.

2. Detecting WIMPS in the Lab

If the halo of our Galaxy is composed of WIMPs, then millions of these particles are
streaming through a square centimeter every second. Goodman and Witten (1985)
and Wasserman (1986) realized that this flux could be experimentally detectable.
There are, however, two difficulties involved with detecting WIMPs:

(1) *They Don't Do Much.* All of the proposed WIMPs have some conserved quantum
number (R-parity for SUSY particles and fourth generation lepton number for
massive neutrinos); hence, the end-product of an interaction with a nucleus is
at best the deposition of a few keV of energy.

(2) *They Don't Do It Often.* The lowest mass supersymmetric particle (which is
stable in many theories) is usually some linear combination of photino and hig-
gsino interaction eigenstates. This Majorana particle has only axial couplings
with quarks and thus has a typical elastic nuclear cross-section $\sim 10^{-37}$ cm^2 for
WIMPs through its spin-dependent interactions with nuclei. If these "sparticles"
were the galactic missing mass, they would produce $\sim 10^{-1} - 1$ counts per day
in a kilogram of detector. Scalar neutrinos and massive Dirac neutrinos have
vector couplings with quarks, and thus the neutrons in the nucleus construc-
tively interfere to yield a much larger cross-section $\sim 10^{-34}$ cm^2 and a higher
count rate $\sim 10^3$ counts/kg/day. In either case, the search for WIMPs requires
a detector with very good background rejection and a low energy threshold.

We already have limits on weakly interacting halo dark matter from germanium
spectrometers. These limits, however, only exclude particles more massive than
16 GeV and it does not seem likely that these limits could be extended below
8 GeV (Ahlen et al., 1987). Unfortunately, the most interesting region for Dirac
neutrinos is around 2–3 GeV (the range of the Lee–Weinberg limit). Particles with
spin-dependent interactions (e.g., photinos, higgsinos, and Majorana neutrinos) also

evade detection since most of the abundant isotopes of germanium have zero spin. (They have an even number of both protons and neutrons.)

This motivates the use of new types of detectors in the search for dark matter. There are several schemes for direct detection of dark matter. One scheme relies on the principle that "simple is beautiful": the use of a single kilogram-mass crystal of silicon to detect phonons produced by WIMP-nucleon scattering (Cabrerra et al., 1985, Sadoulet 1987, Marthoff 1987). There are several groups actively exploring the possibilities of phonon detectors: Cabrerra and Sadoulet and their collaborators at Stanford and Berkeley; Smith and coworkers in Oxford; Moseley and collaborators at NASA.

Other groups guided by the philosophy of "small is beautiful" are focusing on superconducting grains. Here, we focus on these grains which detect the change in state due to the nuclear recoil of a WIMP (Drukier and Vallette 1972, Drukier et al., 1975, Drukier and Stodolsky 1984, Gonzalez-Mestres and Perret-Gallix 1985, Drukier 1987). Several groups are now actively working on developing grains and grain detectors: A. Drukier (Applied Research Corp.) and his collaborators at UBC, Vancouver; G. Waysand and his collaborators in Orsay, Saclay and Annecy; L. Stodolsky, K. Pretzel and collaborators at MPI für Physik und Astrophysik. There are also several other novel proposals for dark matter detectors (e.g. Lanou et al., 1987).

In a detector, the grains would be kept in a superheated superconducting state. The deposition of a few keV of energy in an elastic scatter of a WIMP off of a nucleus in the grain heats the grain and flips it from the superconducting to the normal state. The background magnetic field can now permeate the grain. This produces a change in the magnetic flux through a loop surrounding the grains equivalent to the addition of a dipole whose strength is proportional to the product of the grain's cross-sectional area and the background magnetic field. SQUIDs (superconducting quantum interference devices) would be ideal for detecting this small change in flux. The grain could be composed of any type I superconductor. These include gallium and aluminium, both of which are mostly composed of isotopes with an odd number of neutrons, and thus have large cross-sections for particles with spin-dependent couplings. The grains would probably be coated with a dielectric composed of low Z material. This dielectric separates the grains and suppresses diamagnetic interactions with neighboring grains. The energy threshold of a grain, the minimum amount of energy needed to flip the grain, is set by its composition, size, and temperature. Micron size grains are needed to detect photinos of a few GeV.

The major source of background in the detector is the radioactive decay of trace contaminants in the grains and the surrounding dielectric. Most of these decays produce MeV electrons. Since the dielectric is composed of low Z material, these electrons, which lose energy through Coulomb interactions, will deposit most of their energy in the grains. Thus radioactive decays will flip multiple grains, which will produce a large change in flux in the SQUID loop. The scatter of a WIMP off of a grain will flip only a single grain. Uniform grains, which have similar energy thresholds, are needed for this background suppression mechanism to work. Most of the grains must be in the superheated state so that there is little dead material into which the β particle can deposit its energy.

The problem of grain production is presently the greatest challenge in the development of these detectors. Many of the problems that workers in this field have had with the grains are due to irregularities in grain size and shape. Strong fields near corners of non-spherical grains lead to a spread in energy thresholds. Wide hysteresis curves are a symptom of variations in grain properties. At Vancouver, multiple filtered grains has produced promisingly narrow hysteresis curves which suggest that this process eliminates not only large variations in grain size but also removes irregular grains. Lawrence Livermore National Laboratory has produced very high quality grains. Gonzalez-Mestres and his co-workers at Annecy are obtained high quality grains from a French industrial producer.

Another challenge in dark matter grain detector development will be the integration of the SQUID with the low background technologies. Avoiding background signal due to vibrations producing spurious signals in the SQUID loop will require that the detector is vibrationally well insulated from its environment and may necessitate using multiple SQUID loops as a gradiometer.

Since it is possible for both radioactive decays and the elastic scatter of a halo WIMP to flip a single grain, we must find a characteristic of the signal that will allow differentiation from the background. Failing this, any "detection" could be written off as a misunderstanding of the background. Fortunately, the Earth's motion around the Sun provides a significant modulation in the background rate (Drukier, Freese and Spergel 1985).

The Sun is moving around the Galactic center at a velocity of about 250 km/s. The non-dissipative dark matter in the galactic halo, on the other hand, never collapsed into a disc; thus, it is not rapidly rotating. As a result of the Sun's motion, the detector is moving relative to the galactic halo. Since the grains have an energy threshold, the anisotropy in the velocity distribution alters the predicted count rate.

The Earth's motion around the Sun modulates the velocity of the detector relative to the halo. The Earth moves around the Sun with a velocity of 30 km/s. Since the ecliptic is inclined at 62° relative to the galactic plane, only a fraction of this velocity is added to the Sun's motion. In January, when the Sun is in Sagittarius (the location of the Galactic center), the Earth is moving at ~ 235 km/s relative to the halo. (The Earth's motion around the Sun is counter-clockwise, while the Sun's motion around the Galactic center is clockwise.) In July, more energetic particles will stream through the detector when the Earth is moving at ~ 265 km/s relative to the galactic halo. Since the detector has an energy threshold, the flux of more energetic particles produces a higher count rate. Drukier, Freese and Spergel estimate a modulation in the signal of $\sim 12\%$ in a detector sensitive to 20% of the incident flux. This calculation assumes that the galactic halo has an isothermal distribution of velocities.

The existence of this modulation effect does not depend upon details of models of the galactic halo. It only requires that the Earth's velocity relative to the rest frame of the halo changes with time. The assumption that the halo is composed of WIMPs implies that it is non-dissipative; hence, its rest frame should differ from that of the Sun which is composed of baryons which collected in the disc through dissipation. The only other astronomical requirement for the modulation effect is that the Earth moves around the Sun. Seeing the modulation effect, however, requires that grains have a narrow distribution of energy thresholds ($<50\%$). This is yet another incentive for developing more uniform grains.

In conclusion, a promising application of grain detectors is their use in the search for dark matter. The Earth's motion around the Sun produces a significant modulation which can be used to confirm a detection. While hurdles such as the production of uniform spherical grains remain, recent progress, the rewards of detection, and the powerful limits that could be placed on SUSY theories through non-detection will hopefully continue to motivate experimentalists to surmount these problems.

Astronomy is traditionally a science based on observing photons. The subject of cosmions, however, begins with observations of solar neutrinos and looks towards detecting galactic non-baryonic particles. The major discoveries of the past decades were made possible through the development of instrumentation that opened up new parts of the electromagnetic spectrum. The detection of neutrinos from Supernova 1987a in the LMC revealed some of the promise of non-photonic astronomy. Perhaps the next decades will see the discovery of objects not through their photon emissions, but through our observation of other extra-terrestrial messengers.

References

Ahlen, S., Avignone, F. T. III, R. Brodzinsky, R., Drukier, A. K., G. Gelmini, G., and Spergel, D. N. (1987) "Limits on cold dark matter candidates from the ultralow background germanium spectrometer", submitted to *Phys. Lett.*

Bahcall, J. N. (1984) *Astrophys. J.* **287**, 296.

Bahcall, J. N. (1987) This lecture series.

Bahcall, J. N., Huebner, W. F., Lubow, S. H., Parker, P. D., and Ulrich, R. K. (1982) "Standard solar models and the uncertainties in the predicted capture rates of solar neutrinos", *Rev. Mod. Phys.* **54**, 767.

Bahcall, J. N. and Ulrich, R. K. (1987) *Rev. Mod. Phys.*, submitted.

Bouquet, A. and Salati, P. (1987) "Life and death of cosmions in stars", LAPP-TH-192/87.

Caberra, B., Krauss, L. M., and Wilczek, F. (1985) *Phys. Rev. Lett.* **55**, 25.

Dappen, W., Gilliland, R. L., and Christensen-Dalsgaard, J. (1986) *Nature* **321**, 229.

Drukier, A. K. (1987) "Cryogenic detectors of cold dark matter candidates: some material considerations", IAS Preprint-87-0206, submitted to *Nucl. Instrum. Methods.*

Drukier, A. K., Freese, K., and Spergel, D. N. (1986) "Detecting cold dark matter candidates", *Phys. Rev.*, **D33**, 3495.

Drukier, A. K. and Valette, C. (1972) *NIM*, **105**, 285.

Drukier, A. K. and Stodolsky, L. (1984) *Phys. Rev.*, **D30**, 2295.

Faulkner, J. and Gilliland, R. L. (1985) *Astrophys. J.*, **299**, 994.

Faulkner, J., Gough, D. O., and Vahia, M. N. (1986) *Nature*, **321**, 226.

Faulkner, J. and Spergel, D. N. (1987) "WIMPs and horizontal branch stars", in preparation.

Freese, K. (1986) "Can scalar neutrinos or massive Dirac neutrinos be the missing mass?", *Phys. Lett.*, **B167**, 295.

Gaisser, T. K., Steigman, G., and Tilav, S. (1986) "Limits on cold dark matter candidates from deep underground detectors" *Phys. Rev.*, **D34**, 2206.

Gelmini, G. B., Hall, L. J., and Lin, M. J. (1987) "What is the cosmion?" *Nucl. Phys.*, **B281**, 726.

Gilliland, R., Faulkner, W., Press, W. H., and Spergel, D. N. (1986) "Modelling the effects of weakly-interacting, massive particles on the solar interior", *Astrophys. J.*, **306**, 703.

Goodman, M. and Witten, E. (1985) *Phys. Rev.*, **D31**, 3059.

Gonzalez-Mestres, L. D., and Perret-Gallix, D. (1985) "Detection of magnetic monopoles with superheated type I superconductors", LAPP-EXP-85-02.

Gould, A. (1987a) "WIMP distribution in and evaporation from the Sun", SLAC preprint SLAC-PUB-4162, submitted to *Astrophys. J.*

Gould, A. (1987b) "Resonant enhancements in WIMP capture by the Earth", SLAC preprint, SLAC-PUB-4226, submitted to *Astrophys. J.*

Griest, K. (1987) "Stable, heavy, neutral particles in the Sun and in toponium decay", UC, Santa Cruz SCIPP-87/92, Ph.D. Thesis.

Griest, K. and Seckel, D. (1987) "Cosmic asymmetry, neutrinos, and the Sun", *Nucl. Phys.*, **B283**, 681.

Hagelin, J. S., Ng, K. W., and Olive, K. A. (1986) "A high-energy neutrino signature from supersymmetric relics", *Phys. Lett.*, **B180**, 375.

Hegyi, D. and Olive, K. (1987). "A case against baryons in galactic halos", *Astrophys. J.*, **303**, 56.

Krauss, L. M., Freese, K., Spergel, D. N., and Press, W. H. (1985) "Cold dark matter candidates and the solar neutrino problem", *Astrophys. J.*, **299**, 1001.

Krauss, L. M., Srednicki, M., and Wilczek K. (1986) "Solar system constraints and signatures for dark matter candidates", *Phys. Rev.*, **D33**, 2079.

Lanou, R. E., Maris, H. J., and Seidel, G. M. (1987) "Detection of solar neutrinos in superfluid helium", *Phys. Rev. Lett.*, **58**, 2498.

Marthoff, C. J. (1987) "Limits on sensitivity in large silicon bolometers for solar neutrino detection", *Science*, **237**, 507.

Nauenberg, M. (1986) "Energy transport and evaporation of weakly interacting particles in the Sun", UC, Santa Cruz preprint, SCIPP-87/79.

Primack, J. R. (1986) "Particle dark matter", invited talk given at *2nd ESO/CERN Symp. on Cosmology, Astronomy and Fundamental Physics*, Garching, West Germany, March 17-21, 1986.

Press, W. H. and Spergel, D. N. (1985) "Capture by the Sun of a galactic population of weakly-interacting, massive particles", *Astrophys. J.*, **296**, 663.

Raby, S. R. and West, G. (1987a) "A simple solution to the solar neutrino and missing mass problems", Los Alamos preprint, LA-UR-86-4151, submitted to *Nucl. Phys. B.*

Raby, S. R. and West, G. (1987b) "Experimental consequences and constraints for magninos", Los Alamos preprint, LA-UR-87-1734, submitted to *Phys. Lett. B.*

Renzini, A. (1986) "Effect of cosmions in the Sun and in globular clusters", *Astron. Astrophys.*, **171**, 121.

Sadoulet, B. (1987) "Prospects for detecting dark matter particles by elastic scattering", LBL preprint, LBL-23098, Presented at *13th Texas Symp. on Relativistic Astrophysics*, Chicago, IL, Dec 14-19, 1986.

Silk, J., Olive, K., and Srednicki, M. (1985) "The photino, the Sun and high-energy neutrinos", *Phys. Rev. Lett.*, **55**, 257.

Silk, J. and Srednicki, M. (1984) "Cosmic ray anti-protons as a probe of a photino dominated Universe", *Phys. Rev. Lett.*, **53**, 624.

Spergel, D. N. and Press, W. H. (1985) "Effect of hypothetical, weakly-interacting, massive particles on energy transport in the solar interior", *Astrophys. J.*, **294**, 679.

Srednicki, M., Olive, K. A., and Silk, J. (1987) "High-energy neutrinos from the Sun and cold dark matter", *Nucl. Phys.*, **B279**, 804.

Stecker, F. W., Rudaz, S., and Walsh, T. F. (1985) "Galactic anti-protons from photinos", *Phys. Rev. Lett.*, **55**, 2622.

Steigman, G., Sarazin, C. L., Quintana H., and Faulkner, J. (1978) *Astrophys. J.*, **83**, 1050.

Wasserman, I. (1986) "On the possibility of detecting heavy neutral fermions in the Galaxy", *Phys. Rev.*, **D33**, 2071.

Yang, J., Turner, M., Steigman, G., Schramm, D. N., and Olive, K. (1984) *Astrophys. J.*, **281**, 493.

Chapter 10

AN INTRODUCTION TO COSMIC STRINGS

William H. Press[*]

Harvard-Smithsonian Center for Astrophysics
Cambridge, MA 02138, USA

David N. Spergel[†]

Institute for Advanced Study
Princeton, NJ 08540, USA

This lecture reviews the formation of cosmic strings, the evolution of the cosmic string network, and the behavior of free loops. Also explored is the theory that cosmic string loops are the seeds for galaxy formation.

1. Birth of Cosmic Strings

Cosmic strings sometimes get confused with superstrings. It is, however, easy to distinguish between the two objects: superstrings are a theory of everything that explains nothing in astrophysics, while cosmic strings are a theory of nothing, but have been evoked to explain nearly everything in astrophysics. (This joke seemed funny in the original lecture.)

A physical analog of a cosmic string can be formed in the laboratory. Start with a block of metal in a magnetic field. Slowly cool the block. The outer layer will become superconducting early, expelling magnetic field lines by the Meissner effect. Some magnetic field lines will be trapped inside the block. As more of the metal becomes superconducting, the field lines will be forced into a smaller region. Eventually, they will form a vortex tube — a filament of the normal state trapped inside the superconducting metal. The topological constraint that field lines cannot end ($\nabla \cdot \mathbf{B} = 0$) forces a filament of the metal to stay in the normal state.

Nielsen and Olesen (1974) realized that these vortex tubes (or strings) could form not only at the superconducting phase transition in the laboratory, but also at phase transitions in the early universe. All that is needed is the right sort of phase transition.

Our discussion of inflation (see previous paper) assumed that when the high temperature vacuum state is broken by a spontaneous symmetry breaking, the low temperature vacuum state was unique. However this need not be true, and there

---

[*]Current address: Los Alamos National Lab., DIR/MSA-121, Los Alamos, NM 87545, USA.
[†]Current address: Princeton University Observatory, Princeton, NJ 08544, USA.

may be many possible vacuums. For example, consider a theory in which a Higgs field ϕ couples through a potential of the form $V(\phi) = \lambda(\phi^2 - \eta^2)^2$. If ϕ is real, then there are two discrete vacuum states: $\phi = \eta$ and $\phi = -\eta$. This discrete two-fold symmetry is called a Z_2 symmetry and will not produce cosmic strings, but rather another class of topological defects called domain walls.

When a phase transition occurs in the early universe, the gradient term $(D^\mu \phi D_\mu \phi)$ in the Lagrangian will couple neighboring regions of space and encourage them to assume the same vacuum value. Casually disconnected regions; however, cannot communicate and can assume different vacuum states. If in one region, ϕ has a vacuum expectation value of η and in neighboring region, ϕ has a vacuum expectation value of $-\eta$, then there must be an intervening region were ϕ passes through 0, and the original false vacuum state is restored. This narrow surface that separates two regions with different vacuum expectation values is called a domain wall. Domain walls are of little interest to most cosmologists, since if they did exist they would have been observed as the dominant form of matter/energy in the universe.

Cosmic strings require a more complicated theory. If ϕ is complex, then its vacuum state will have a U(1) symmetry: $\phi = \eta e^{i\theta}$. At each point in space the field can assume a phase, θ, between $0°$ and $360°$. The gradient term will try to line up the phases, however, it is possible to have a region in which the phase around a loop changes by $360°$, this implies the existence of a critical point at which the phase is ill defined.

Brandenberger (1987) has offered a simple proof of why these points must connect to form strings. Consider a closed curve. If the integral of $\nabla \theta$ around a path is 2π, then there must be a singularity somewhere on the surface inclosed by the curve. $\nabla \theta$ can only be singular when $\phi = 0$. Thus somewhere on this surface, there must be a point in the false vacuum state. We can deform this surface and find yet another point of false vacuum. These points connect together to form a line.

The scale of the symmetry breaking, η, that produces the string network determines the mass per unit of length of the string, $\mu \approx \eta^2/M_{PL}$, where $M_{PL} = c^2/G$ is the Planck scale (10^{19} GeV). This mass per unit length can be enormous,

$$\mu \approx 10^{28} \left(\frac{G\mu}{c^2} \right) \text{g/cm}^3 \approx 10^{13} \left(\frac{G\mu}{c^2} \right) M_\odot/\text{pc} . \tag{1}$$

In typical GUT theories, $\eta \approx 10^{16}$ GeV and $G\mu/c^2 \approx 10^{-6}$. The last section will review how these massive strings may seed galaxy formation.

An unambiguous numerical scheme that simulates a string-generating phase transition is to carve up a volume of space into a lattice of tetrahedrons and randomly assign 1,2, or 3 to each vertex. Strings enter the tetrahedrons through cyclic faces (123) and exit through an anti-cyclic faces (321). This procedure of imposing of Z_3 symmetry on the tetrahedron lattice avoids the ambiguity of schemes that use cubes.

Simulations of phase transitions have suggested that strings either reconnect quickly or wander off to infinity has part of an infinite lattice (Kibble 1976, Vachaspati and Vilenkin 1984). Approximately 60% of the length of the loops is in the infinite length string; the remaining length is in small loops with a distribution $dn \propto l^{-2.5} \, dl$. Scherrer and Frieman (1987) have explored the properties of the infinite string network and have found that it behaves not like a Brownian random walk, but rather a self-avoiding random walk, as observed in a dilute solution of polymers.

The string network will oscillate and strings will cross. Three-dimensional simulations of string crossing suggest that the probability of "intercommutation" is high. This "intercommutation" is akin to the reconnection of field lines in an MHD plasma. This probability is certainly a function of angle and velocity; however, in simulations of string evolution, it is usually assumed to be unity.

As the horizon expands, the infinite string network reconnects and forms loops. The largest scale at which loops can form is the horizon size. There is some numerical and theoretical evidence that this process of loop formation is self-similar (Albrecht and Turok 1985, Bennett 1986): there is only one infinite loop crossing the horizon at any time, the rest of the string's length is in closed loops. This suggest that the mass in strings with in the horizon is $\approx K\mu ct$, where K is a constant and ct is the length of the infinite loop that cross the horizon. The volume within the horizon is growing as $c^3 t^3$, while the energy density in radiation decreases, $\rho_{\text{rad}} \sim aT^4 \sim 1/Gt^2$. This suggests that the energy density in strings is a constant fraction of the radiation energy density:

$$\frac{\rho_s}{\rho_{\text{rad}}} \approx \frac{K\mu ct/(ct)^3}{1/Gt^2} \approx \left(\frac{G\mu}{c^2}\right) K \approx 10^2 \frac{G\mu}{c^2}. \tag{2}$$

This solution might possibly not be correct, however. It is possible that the probability of an infinite string fragmenting into a infinite string plus loop is the same as that of a loop rejoining an infinite string. This would allow the string network to continually reassemble itself. Since the length of string per comoving volume $\propto R$, the energy density in strings would scale as R^{-2}. Strings would very rapidly dominate the universe.

A critical universe dominated by strings has the same dynamics as empty flat space. The string's negative pressure, $p = -\frac{1}{3}\rho$ cancels its density in the expansion equation:

$$\frac{\ddot{R}}{R} = \frac{4\pi G}{3}(\rho + 3p) = 0. \tag{3}$$

The string dominated universe expands linearly.

Most of the interest in strings have focused not on the string-dominated universe, but rather on a matter-dominated universe in which string loops are the seeds for the formation of galaxy via gravitational collapse. The rest of the lecture will discuss the dynamics of these string loops.

2. The Motion of a Cosmic String Loop

We will derive the string's equations of motion by analogy with a rubber band. We can label points along the rubber band with s. The function $\vec{x}(s,t)$ describes the position of points along the rubber band at all times. The rubber band's equation of motion can be derived from its action,

$$S = \int \int \mathcal{L} ds \, dt, \tag{4}$$

where the Lagrangian has a kinetic energy and a tension contribution:

$$\mathcal{L} = \frac{1}{2}\mu\dot{x}^2 - \frac{1}{2}Tx'^2 . \tag{5}$$

Dot denotes a time derivative and prime denotes a derivative with respect to s.

The Euler-Lagrange equation,

$$-\frac{\delta L}{\delta \vec{x}} + \frac{\partial}{\partial t}\frac{\delta L}{\delta \dot{\vec{x}}} + \frac{\partial}{\partial s}\frac{\delta L}{\delta x'} = 0 \tag{6}$$

yields for constant μ:

$$\ddot{\vec{x}} = \frac{T}{\mu}\vec{x}'' \tag{7}$$

which is instantly recognizable as a wave equation that separates for each Cartesian component of $\vec{x}$. We now multiply the string equation of motion by $\dot{x}$ and use the identity $\frac{d}{ds}(\vec{x}' \cdot \dot{\vec{x}}) = \vec{x}'' \cdot \dot{\vec{x}} + \vec{x}' \cdot \dot{\vec{x}}'$ to obtain

$$\frac{1}{2}\frac{d}{dt}(\dot{\vec{x}}^2) = \frac{T}{\mu}\left(\frac{1}{2}\frac{d}{dt}(\vec{x}')^2 + \frac{d}{ds}(\vec{x}' \cdot \dot{\vec{x}})\right). \tag{8}$$

If we integrate the above equation around the rubber band, we obtain a equation of total conservation of energy:

$$\frac{d}{dt}\left[\int ds\left(\frac{1}{2}\dot{\vec{x}}^2 + \frac{T}{\mu}(\vec{x}')^2\right)\right] = 0. \tag{9}$$

This suggests a reparamaterization of labels along the string so that we are tracking energy packets rather than bits of string. We can make the gauge choice $\vec{x}' \cdot \dot{\vec{x}} = 0$ which leads to a local energy conservation law (see Eq. 8).

We now turn to a cosmic string. The action for a cosmic string,

$$S = \mu \int ds \, d\tau \, \mathcal{L} \tag{10}$$

is the proper area of the world sheet swept out by the string. Recall that the action of a free point particle is the length of its world line. The area of a world sheet defined by $\vec{x}'$ and $\dot{\vec{x}}$ is:

$$\mathcal{L} = |\vec{x}'||\dot{\vec{x}}|\sin\theta = \sqrt{\dot{\vec{x}}^2\vec{x}'^2 + (\dot{\vec{x}} \cdot \vec{x}')^2}. \tag{11}$$

The action associated with the area of a world sheet is called the Nambu action. It has attracted much attention in recent years as the action for superstrings. In superstring theory the strings are moving in a 10 dimensional space rather than in the more familiar 4 dimensional space. This action has a property called Weyl invariance: The action is invariant under any conformal transformation. This new symmetry can have powerful mathematical and physical implications.

Written (1985) demonstrated that in certain particle theories, cosmic strings can carry currents and behave much like superconducting wires. The addition of electromagnetic interactions enriches the physics of the cosmic string by adding new terms to the string action. The string can radiate not only graviational waves, but also electromagnetic radiation. This lecture will concentrate on non-conducting strings. We refer the interested reader to Spergel, Piran and Goodman (1987) for a detailed treatment of the properties of superconducting cosmic strings.

When a non-conducting string loop is much smaller than the horizon size, it obeys a simple wave equation that is derived from Eq. (11) in flat space:

$$\ddot{\vec{\mathbf{x}}} = \vec{\mathbf{x}}'' \tag{12}$$

in the gauge in which $x^0 = t, \vec{x}' \cdot \dot{\vec{x}} = 0, (\vec{x}')^2 + (\dot{\vec{x}})^2 = 0$. Turok (1984) emphasized that the string's oscillations are the sum of left and right moving waves:

$$\vec{\mathbf{x}} = \frac{1}{2}\big[\vec{\mathbf{a}}(t + s) + \vec{\mathbf{b}}(t - s)\big]. \tag{13}$$

The gauge conditions,

$$\dot{\vec{x}} \cdot \vec{x}' = \frac{1}{4}\big[(\dot{\vec{\mathbf{a}}} + \dot{\vec{\mathbf{b}}}) \cdot (\dot{\vec{\mathbf{a}}} - \dot{\vec{\mathbf{b}}})\big] = \frac{1}{4}(\dot{\vec{\mathbf{a}}}^2 - \dot{\vec{\mathbf{b}}}^2),$$

$$\dot{\vec{x}}^2 + \vec{x}'^2 = -1 + \frac{1}{4}\big[2\dot{\mathbf{a}}^2 + 2\dot{\mathbf{b}}^2\big], \tag{14}$$

imply that $|\dot{\vec{\mathbf{a}}}| = |\dot{\vec{\mathbf{b}}}| = 1$. These two vectors describe paths on the unit circle. Since $\dot{\vec{\mathbf{a}}}$ and $\dot{\vec{\mathbf{b}}}$ are derivatives of periodic functions, their mean must be zero. Thus they will visit every hemisphere. This implies that they will generically cross. When they cross, $\dot{\vec{\mathbf{a}}} = \dot{\vec{\mathbf{b}}}$, and a piece of the string reaches the speed of light. The point is called a cusp at it is a square-root singularity in $\vec{x}'$. Because of their high velocities, cusps are the dominant of gravitational and electromagnetic radiation from the string loop.

Not all string loops must have cusps. There are special solutions without crossings. For example, $\vec{a}$ and $\vec{b}$ can follow the "seams of a baseball" and avoid cusps. Alternatively, there can be discontinuities in either curve. Reconnections naturally produce these discontinuities which allow one curve to "jump" across the other and avoid forming a cusp. These discontinuities result in kinks that propagate along the string. Kinks are points at which the string's velocity changes discontinuously. It is not known whether gravitational and/or electromagnetic radiation from the string will dampen these kinks and restore cusps. Recent articles by Vachaspati and Garfinkle (1987) and Thompson (1987) provide lucid discussions of recent work on kinks and cusps.

Oscillating string loops can cross themselves reconnect and split into smaller loops. Self-intersections can be found by examining $\dot{\vec{a}}$ and $\dot{\vec{b}}$, rather than $\ddot{\vec{a}}$ and $\ddot{\vec{b}}$. Considering the curves $\vec{a}(\xi)$ and $\vec{b}(\xi)$ allows us to either follow a point, $\vec{x}(t) = \vec{a}(\xi) + \vec{b}(\xi)$, or to trace the whole string at a point in time, $\vec{x}(s) = \vec{a}(\xi) + \vec{b}(-\xi)$.

Crossings will occur whenever there is a pair of points along the string (s_1, s_2) such that,

$$\vec{a}(t + s_1) + \vec{b}(t + s_1) = \vec{a}(t + s_2) + \vec{b}(t + s_2). \tag{15}$$

This will happen whenever a pair of chords on the $\vec{a}$ and $\vec{b}$ curves inscribe the same arc-length and are parallel and of equal length, a point emphasized by R. Scherrer,

$$\vec{a}(t + s_1) - \vec{a}(t + s_2) = \vec{b}(t + s_1) - \vec{b}(t + s_2). \tag{16}$$

While searching N points on the $\dot{\vec{a}}$ and $\ddot{\vec{a}}$ curves for cusps in a N^2 process, examining all pairs of curves for self-intersections is an N^3 search. Since there is both an extra dimension in the search for crossings and an additional constraint, self-intersections are also generic.

Witten has remarked that it is a special property of strings in a 4 dimensional universe that both cusps and self-intersections are generic. In a higher dimensional space, cusps are a set of measure zero. In a lower dimensional space, the set of non-intersecting loops in a set of measure zero. This result is linked to the fact that two dimensional surfaces in four space generically cross.

In an expanding universe, strings behave according to a modified equation of motion. The Lagrangian can be rewritten for the Robertson-Walker metric,

$$ds^2 = -dt^2 + R(t)^2(dr^2 r^2 d\Omega^2)$$

in the gauge in which $\dot{\vec{x}} \cdot \vec{x}' = 0$ and $dx^0/dt = 1$:

$$\mathcal{L} = R^2 \left(\frac{d\vec{x}}{ds}\right)^2 \left(-1 + R^2 \left(\frac{d\vec{x}}{dt}\right)^2\right). \tag{17}$$

The Euler–Lagrange equations now yield a more complicated equation of motion:

$$\frac{d}{dt}\left(R^2\dot{\vec{x}}\sqrt{\frac{R^2(\vec{x}')^2}{1 - R^2(\dot{\vec{x}})^2}}\right) = \frac{d}{ds}\left(R^2\vec{x}'\sqrt{\frac{R^2(\dot{\vec{x}})^2}{1 - R^2(\vec{x}')^2}}\right). \tag{18}$$

By defining an effective energy per unit length,

$$\epsilon = \sqrt{R^2(\vec{x}')^2/(1 - R^2(\dot{\vec{x}})^2)},$$

Eq. (18) can be simplified to

$$\frac{d}{dt}(R^3\dot{\vec{x}}\epsilon) = R\frac{d}{ds}\left(\frac{\vec{x}'}{\epsilon}\right). \tag{19}$$

We can find an energy equation for a string oscillating in an expanding universe by multiplying by $R\epsilon\dot{\mathbf{x}}$ and integrating,

$$\dot{\epsilon} = -\frac{2\dot{R}}{R}\epsilon R^2(\dot{\mathbf{x}})^2. \tag{20}$$

The energy loss is due to the Doppler shifting of the loops' velocity as the Universe expands.

If the stings do not intersect, they will continue along the same trajectories, returning to the same configuration every oscillation. These oscillating strings have variable quadrupole moments, thus are a source of gravitational radiation. The graviational radiation from an oscillating source,

$$L_{\text{Grav}} \approx \frac{L_{\text{int}}^2}{c^5/G}, \tag{21}$$

where L_{int} is the internal luminosity of the string. A string of length R has a mass of $\sim\mu R$. Every oscillation period, $R/2c$, the string moves $\mu c^2 R$ of energy through space. The string's internal luminosity is $E/t - \mu c^3$, thus it radiates,

$$L_{\text{Grav}} \approx \left(\frac{G\mu}{c^2}\right)\mu c^3 \tag{22}$$

in gravitational radiation. This implies that a string loses $G\mu/c^2$ of its length every oscillation time. Ostriker, Thompson and Witten (1986) suggest that if the string carries electric current, it can emit a comparable amount of electromagnetic radiation.

3. Cosmic Strings and the Formation of Galaxies

Over the past few years, there has been growing interest in the hypothesis that cosmic strings could seed the growth of galaxies in the early universe. Initial work suggests that cosmic strings could account for the number and distribution of clusters and galaxies and explain the slope and amplitude of the cluster correlation function (Zeldovich 1980, Vilenkin 1981, Brandenberger and Turok 1986 and references therein). Peebles (1986a, 1986b) has alerted workers in the field to discrepancies between observations and the predictions of the cosmic string galaxy formation scenario. Peebles claims that the cosmic string scenario does not explain the large scale topology seen by deLapparent et al. (1986), nor does it explain large scale biasing. Vachaspati (1987) has suggested that wakes of infinite string loops may be responsible for the large scale frothy structure. Stebbins et al. (1987) claims that these wakes may alleviate some of Peebles' concerns.

Peebles (1986b) also questions whether the objects that accrete around string loops would look like galaxies. The spherical accretion model yields structures with small core radii and predicts too steep a slope in the galaxy luminosity function.

These discrepancies motivate a reconsideration of how matter accretes around cosmic string loops. We will focus on how competition between loops limits their accretion of dark matter and baryons and the role that galactic cannabalism may play in the evolution of galaxies in the cosmic string scenario.

One of the challenges in studying the astrophysical implications of these massive cosmic string loops is understanding how to associate galaxies and clusters with these loops. Scherrer and Melott (1987) argue against the $1:1$ association of large cosmic string loops with rich Abell clusters. Their numerical simulations of cosmic string seeded galaxy formation do not reproduce the Bahcall and Soniera (1983) cluster-cluster correlation function.

The next section will review spherical accretion model of Gott (1975) and Gunn (1977). This model ignores the effects of neighboring loops on the accretion of matter and the motion of cosmic string loops. Bertschinger (1987) has shown that the string's motion does not affect the amount of matter accreted around the loop, only the shape of the galaxy. We will then consider the competition between loops and show that the amount of mass that accretes onto a loop is initially proportional to its length. This implies that initially most of the matter in a group or cluster is accumulated in small $\sim 10^{-2}L_*$ objects. Larger systems may cannabalize these smaller objects and steepen the correlation between galaxy mass, M_g, and loop radius, R_L to $M_g \propto R_L^{3/2}$. If L_* galaxies can cannabalize their smaller companions, then the cosmic string model successfully predicts the slope of the galaxy luminosity function and the dependance of luminosity on velocity dispersion ["Faber-Jackson Law" (Faber and Jackson 1976)]. If cannabalism is an inneffecent process, then observed galactic morphology conflicts with theoretical expectations.

3.1. *Spherical Accretion Model*

We review the accretion of cold dark matter around a loop of radius R and mass, $M = \beta\mu R$, focussing on the "standard" scenario of Brandenberger and Turok (1986). Zurek (1986) has suggested that a cascade of self-intersections would produce a dramatically different morphology. Bertschinger and Watts (1987) explore the accretion of neutrinos ("hot" dark matter) onto cosmic string loops.

Albrecht and Turok (1985) find that their numerical simulations suggest a self-similar spectrum of string loops. The number density of loops, $n(R)$, of radius R, produced in the radiation epoch ($R < ct_{eq}$):

$$n(R,t)dR = \nu R^{-5/2}t_{eq}^{1/2}c^{3/2}t^{-2} \tag{23}$$

where $\nu \approx 0.01$ and $t_{eq} = 5.4 \times 10^{10}\,h^2$ sec in a universe with 3 light neutrinos species. These loops oscillate and emit gravitational radiation. Vachaspati and Vilenkin (1984) have estimated that the luminosity of a typical loop is $\sim 50G\mu^2 c$. This implies that the loop distribution function is cut off at $R_{min}(t) = 50(G\mu/c^2)ct$. Loops that were formed with smaller initial radii have decayed away through gravitational radiation losses.

The typical distance to the nearest loop of radius greater than R can be determined from Eq. (23):

$$d(R) = \left(\frac{9}{8\pi\nu} \right)^{1/3} R^{1/2} t_{eq}^{-1/6} t^{2/3}$$

$$= 5.5 \left(\frac{R}{1\,\text{pc}} \right)^{1/2} h^{-4/3}(1+z)^{-1}\text{Mpc}. \tag{24}$$

Since $d(R) \gg R$, accretion around a loop can be modelled as accretion around a point mass of mass $\beta\mu R$. Gott (1975) and Gunn (1977) have considered spherical accretion in a cosmological context. Brandenberger and Turok (1986) applied this model to galaxy formation around cosmic string loops.

Density perturbation will grow around all loops that survive to the matter-dominated epoch, this implies that the smallest seed loops will be 0.05 pc and the spacing between loops will be 1 Mpc.

Consider a spherical shell of radius r_i, centered around the cosmic string loop initially has Hubble velocity $H_i r_i$. In a radiation dominated universe, the rapid expansion of the universe prevents the cosmic string loop from having a substantial effect on the motion of the shell. [See Stebbins (1985) for discussion of the growth of the pertubation in the radiation epoch.] After t_{eq}, the excess density due to the string loop has the effect of slowing the expansion of the shell. The shell that started at distance r_{eq} from the string at time t_{eq} will expand until a time,

$$t_{max} = \frac{3\pi}{4} \left(\frac{2}{9G\mu\beta R} \right)^{3/2} r_{eq}^{9/2} t_{eq}^{-2} c^{-3}, \tag{25}$$

at which it has expanded to r_{max}:

$$r_{max} = \frac{4\pi r_{eq}^4}{3\beta\mu Rc^2 t_{eq}^2}$$

$$= \left(\frac{4}{3\pi} \right)^{8/9} \left(\frac{9\beta G\mu R}{2} \right)^{1/3} t_{eq}^{-2/9} t^{8/9}. \tag{26}$$

This shell will then collapse and the material in the shell will have a final radius of $\sim 0.5 r_{max}$. This accretion will build up a density profile around the loop,

$$\rho(R,r) \simeq \frac{1}{3} \left(\frac{3\beta}{2\pi} \right)^{3/4} \rho_{eq}^{1/4} (R\mu)^{3/4} r^{-9/4} \tag{27}$$

where r is the distance from the galactic center.

We can predict the rotation velocity, $v(r)$ by equating the centrifugal force v^2/r to the gravitational force, $GM(r)/r^2$. This yields a nearly flat rotation curve:

$$v(r) = \pi G \left(\frac{3\beta}{2\pi} \right)^{3/8} \rho_{eq}^{1/8} (R\mu)^{3/8} r^{-1/8}. \tag{28}$$

Baryonic infall; however, will steepen the rotation curve, possibly producing a disagreement with observations (Peebles 1986a).

The spherical accretion model ignores competition from other loops. We can estimate the importance of competition by comparing the maximum radius reached by a shell turning around a time t around a loop of radius R with the distance to the nearest loop of radius R or larger:

$$\frac{r_{\max}(t)}{d(R,t)} = 4\left(\frac{R}{1\mathrm{pc}}\right)^{-1/6}(1+z)^{-1/3}\left(\frac{\mu}{2\times10^{-6}}\right)^{1/3}h^{-1/3}. \tag{29}$$

This ratio is greater than 1 for loops that will form L_* galaxies: all of the matter has already accreted into galaxy-size objects by a redshift of 1. This implies that the total mass accreted onto a loop is limited by competition with loops of comparable size. The big loops also must compete with the smaller loops for baryons and dark matter. We now consider this competition.

3.2. Competition between Loops

Consider two loops a distance $s_0/(1+z)$ apart. The longer loop has radius, R_L, while the smaller loop of radius has radius, R_s. Balancing the tidal force due to the larger loop against the gravitational force of the large loop implies that all material inside a radius $r_t(R_s, R_L)$ is bound to the smaller loop where,

$$r_t(R_s, R_L) = (R_s/2R_L)^{1/3}s_0/(1+z). \tag{30}$$

The total mass accreted by the smaller loop, M_s, is determined by its distance from the larger loop:

$$M_s = \frac{2\pi}{3}\left(\frac{R_s}{R_L}\right)\rho_0 s_0^3, \tag{31}$$

where ρ_0 is the current matter density.

In the entire shell of radius s around the larger loop, the fraction f of material bound to smaller loops,

$$f(s_0, R_L) = \rho_0^{-1}\int_{R_{\min}(t_{\mathrm{eq}})}^{R_L} n(R)M_s(R_s, R_L, s_0)dR \tag{32}$$

can be calculated from Eqs. (23), (30), and (31):

$$f(s_0, R_L) = \frac{4\pi\nu}{3}\frac{(ct_{\mathrm{eq}})^{1/2}s_0^3}{c^2t_0^2R_L}\left(R_{\min}^{-1/2} - R_L^{-1/2}\right) \tag{33}$$

where t_0 is the current age of the universe. There is a critical distance, s_{crit}, beyond which $f(s) > 1/2$ and most of the material is bound to smaller loops. The loops that form L_* galaxies have radii much larger than $R_{\min} = \gamma G \mu c t_{\text{eq}}$, thus

$$s_{\text{crit}}(R_L) \approx \left(\frac{3}{8\pi\nu}\right) (ct_0)^{2/3} R_L^{1/3} (\gamma G \mu)^{1/6}. \tag{34}$$

The total mass that will initially bind directly to a large loop, M_{init}, is

$$M_{\text{init}} = \frac{\rho_0(ct_0)^2 R_L}{2\nu}(\gamma G \mu)^{1/2}. \tag{35}$$

This implies that the mass initially accreted directly onto a loop is proportional to its length. Since most of the length of the string network is bound up in the loops of length $\sim R_{\min}$, most of the mass initially accretes onto dwarf galaxies of mass,

$$M_{\text{sat}} = \frac{4\pi}{3}\rho_0 d(R_{\min}, t_0)^3. \tag{36}$$

3.3. Galaxy Morphology

Now, we attempt to compare the predictions of the spherical accretion model with observations of galaxy properties. If we assume that the mass to light ratio of the accreted matter does not vary systematically with loop radius, then Eq. (23) implies that the galaxy luminosity scales linearly with the radius of the seed loop: $L \propto M \propto R$. Combining this relationship with Eq. (35) yields a very steep galaxy multiplicity function:

$$n(L)dL \propto n(R)dR \propto L^{-5/2}dL. \tag{37}$$

Peebles (1987b) has noted the glaring discrepancy between the large number of dwarf galaxies predicted in the cosmic string scenario and the small number of dwarf galaxies observed in the local group.

Galaxy luminosity scaling linearly with loop radius also does not predict the Faber-Jackson law. Equation (35) implies too steep a velocity-luminosity relationship:

$$v \propto R^{3/8} \propto L^{3/8}. \tag{38}$$

Faber and Jackson's (1976) observations of elliptical galaxies reveal that $v \propto L^{1/4}$.

Galactic cannabalism can perhaps save the cosmic string scenario. The previous section showed that the matter accreted onto most of the smaller loops were bound to the large loops. These subsystems may have merged. These systems have had several dynamic times to interact. In small groups, galaxy mergering is a rapid and efficent process. Perhaps, galaxies should not be identified with one string loop, but rather with the material that accreted around a collection of seed loops.

If cannabalism is important, then the mass accreted around an individual loop will be determined by the distance to the nearest loop of comparable size (or bigger), $d(R)$. Equation (24) suggests a steeper relation between loop size and galaxy mass:

$$M \propto d(R)^3 \propto R^{3/2}. \tag{39}$$

Cannabalism reduces the number of small galaxies and correctly predicts the Faber-Jackson law:

$$v \propto L^{1/4}. \tag{40}$$

This analysis suggests that N-body simulations of the accretion of matter around a collection of loops are needed to reveal if collective effects will remove the discrepancy between the "standard" cosmic string model and observations.

4. Observing Cosmic Strings

Cosmic strings may be detected in the next few years. There are several possible ways of observing the effects of cosmic string loops.

Cosmic strings can act as gravitational lenses. Gott (1984) and Vilenkin (1984) show that a string crossing a segment of the sky would form pairs of images of equal brightness seperated by $G\mu/c^2$ radians. If $G\mu/c^2 \sim 10^{-6}$, then a cosmic string would act as an arc-second lens. Cowie and Hu (1987) have already detected 4 pairs of galaxies of nearly equal brightness that they suggest as a candidate for lensing by a cosmic string loop. Further observations will test the exciting possibility.

Hogan and Rees (1984) showed that the gravitational radiation from decaying cosmic strings is also detectable. The millisecond pulsar monitored by Taylor and his collaborators is a superb clock. A gravitational wave passing between the Earth and a pulsar would produce a variation in the Earth-pulsar distance and noise in the clock. Current observations rule out $G\mu > 10^{-5}$. As the pulsar is observed over longer time baselines, limits on $G\mu$ will improve with observing time as t^8! If the pulsar behaves itself, we will soon be able to rule out strings massive enough to form galaxies ($G\mu > 10^{-6}$).

In certain theories, cosmic strings can behave as superconductors (Witten 1985). The presence of electromagnetic interactions enables the string loop to have dramatic effects on its environment. Chudnofsky et al. (1986) suggest that synchotron emission from hot plasma interacting with the magnetic fields around superconducting strings could be detected. Hill et al. (1987) and Babul et al. (1987) propose observing ultra-high energy emission emitted from cosmic string cusps. If strings are superconducting, they could be detected even if $G\mu \ll 10^{-6}$.

The detection of a cosmic string loop would be a dramatic discovery. It would reveal the existence of a phase transition at scales inaccessible in the laboratory and would provide a window into the very earliest moments of the universe.

References

Albrecht, A. and Turok, N. (1985) *Phys. Rev. Lett.* **54**, 1868.
Babul, A., Paczynski, B., and Spergel, D. N. (1987) *Astrophys. J. Lett.*, **316**, L49.

Bahcall, N. and Soniera, R. (1983) *Astrophys. J.*, **270**, 20.

Bennett, D. P. (1986) *Phys. Rev.*, **D34**, 3592.

Bertschinger, E. (1987) *Astrophys. J.*, **316**, 489.

Bertschinger, E. and Watts, P. N. (1987) "Galaxy formation with cosmic strings and massive neutrinos", submitted to *Astrophys. J.*

Brandenberger, R. H., Albrecht, A., and Turok, N. (1986) *Nucl. Phys.*, **B277**, 605.

Brandenberger, R. H. (1987) "Inflation and cosmic strings: Tow mechanisms for the formation of large scale structure", DAMTP preprint.

Chudnovsky, E. M., Field, G. B., Spergel, D. N., and Vilenkin, A. (1986) *Phys. Rev.*, **D34**, 944.

Cowie, L. L. and Hu, E. M. (1987) *Astrophys. J.*, **318**, L33.

deLapparent, V., Geller, M., and Huchra, J. (1986) *Astrophys. J. Lett.*, **301**, L1.

Faber, S. M. and Jackson, R. E. (1976) *Astrophys. J.*, **204**, 668.

Gott, J. R. (1975) *Astrophys. J.*, **201**, 297.

Gunn, J. E. (1977) *Astrophys. J.*, **2198**, 592.

Hill, C. T., Schramm, D. N., and Walker, T. P. (1986) "Ultra high-energy cosmic rays from superconducting cosmic strings", FERMILAB preprint PUB-86/146-T.

Hogan, C. J. and Rees, M. J. (1984) *Nature*, **311**, 109.

Kaiser, N. and Stebbins, A. (1984) *Nature*, **310**, 391.

Kibble, T. W. B. (1976) *J. Phys.*, **A9**, 1387.

Kibble, T. W. B. and Turok, N. (1982) *Phys. Lett.*, **116B**, 141.

Melott, A. L. and Scherrer, R. J. (1987) "The formation of large-scale structure from cosmic string loops and cold dark matter", University of Kansas, preprint.

Nielsen, H. and Olesen, P. (1973) *Nucl. Phys.*, **B61**, 45.

Ostriker, J. P., Thompson, C., and Witten, E. (1986) *Phys. Lett.*, **B180**, 231.

Peebles, P. J. E. (1986a) "A screed on the cosmic string scenario for gravitational galaxy formation", private communication.

Peebles, P. J. E. (1986b) "The cosmic string scenario for gravitational galaxy formation: Screed II", private communication.

Scherrer, R. J. and Frieman, J. A. (1986) *Phys. Rev.*, **D33**, 3556.

Spergel, D. N., Piran, T., and Goodman, J. (1987) *Nucl. Phys.*, **291**, 847.

Stebbins, A. (1986) *Astrophys. J. Lett.*, **303**, L21.

Stebbins, A., Brandenberger, R., Veeraraghavan, S., Silk, J., and Turok, N. (1986) "Cosmic string wakes", UC Berkeley, Print-86-1155.

Thompson, C. (1987) preprint.

Turok, N. (1984) *Nucl. Phys.*, **B242**, 520.

Turok, N. and Brandenberger, R. (1986) *Phys. Rev.*, **D33**, 2175.

Vachaspati, T. and Garfinkle, R. (1987) to appear in *Phys. Rev. D*.

Vachaspati, T. and Vilenkin, A. (1984) *Phys. Rev.*, **D30**, 2036.

Vachaspati, T. and Vilenkin, A. (1985) *Phys. Rev.*, **D31**, 3052.

Vilenkin, A. (1981) *Phys. Rev. Lett.*, **46**, 1169.

Vilenkin, A. (1985) *Phys. Rep.*, **121**, 263.

Witten, E. (1985) *Nucl. Phys.*, **B249**, 557.

Zel'dovich, Ya. B. (1980) *Mon. Not. R. Astron. Soc.*, **192**, 663.

Zurek, W. H. (1986) *Phys. Rev. Lett.*, **57**, 2326.

Chapter 11

A DEPARTURE FROM NEWTONIAN DYNAMICS AT LOW ACCELERATIONS AS AN EXPLANATION OF THE MASS-DISCREPANCY IN GALACTIC SYSTEMS

Mordehai Milgrom

Department of Physics, Weizmann Institute
Rehovot 76100, Israel

Outline of the Lectures

1. The dark side of the dark matter hypothesis; wastefulness, arbitrariness, uselessness.
2. All dynamical determinations are based on a single relation between mass, velocity, and separation.
3. A modification of only the r dependence of gravity conflicts with observations.
4. The basic assumptions of MOND:

   a. Departure at low accelerations.

   b. There exists a_o such that for $a \ll a_o$: $a^2/a_o \approx MGr^{-2}$.

5. Is MOND an amendment of the second law or of the law of gravity?
6. MOND is not a modification at large separations (it may also be required when dealing with small systems such as the long-period comet system). MOND entails accelerations that are nonadditive (nonlinear) in the attracting masses.
7. Different theories can be built on the basic assumptions [different forms of $\mu(x)$, theories with more than one potential, etc.].
8. One finds $a_o \sim cH_o$. Mach's principle. Does a_o vary with cosmic time?
9. A nonrelativistic gravitational potential theory: $\vec{\nabla} \cdot \{\mu(|\nabla\varphi|/a_o)\vec{\nabla}\varphi\} = 4\pi G\rho$.

   a. Conservation of momentum, angular momentum, and energy.

   b. A composite (but small and light) particle with arbitrary internal accelerations falls, in an ambient field, as does a test particle.

10. In systems with spherical (or cylindrical, or planar) symmetry $\mu(g/a_o)\vec{g} = \vec{g}_N$.
11. The ambient field effect: The internal dynamics inside a system that is falling in an external field may be strongly affected by the presence of that field. The strong equivalence principle does not hold.
12. The dynamics in a system immersed in a dominant ambient field is quasi-Newtonian but has a preferred direction and an effective gravitational constant larger than G.

Consequences of MOND.

13. Galaxies (in particular discs) hold the key to the mass discrepancy. Their observations involve good measurements, clear interpretation, large samples, well-defined regularities.
14. No need for dark matter in galaxy binaries, small groups, clusters, and the Virgo infall system.
15. Rotation curves: $L(\vec{r}) \to M(\vec{r}) \to v_{\text{MOND}}(r) \overset{?}{=} v_{\text{obs}}(r)$.
16. In MOND, the shape of the rotation curve depends not only on the mass distribution but also on its normalization.
17. The rotation curve of an isolated galaxy is asymptotically flat, $v \to v_\infty$.
18. $v_\infty^4 = MGa_\text{o}$.
19. The surface density $\Sigma_\text{o} \equiv a_\text{o}G^{-1}$ plays a special role.
20. Isothermal spheres as models of ellipticals, galactic bulges, and galaxy clusters:

   a. are finite, $\rho \xrightarrow{r \to \infty} r^{-\alpha}$ $(\alpha > 3)$.
   b. $\Sigma(\equiv M/r_{1/2}^2) \lesssim \Sigma_\text{o}$.
   c. $a_\text{o}GM = Q\sigma^4$; $1 \le Q \le 9/4$.

21. The general expression for the phantom mass density: $\rho(\text{luminous}) \to \vec{g}(\text{MOND}) \to \rho^*(\text{Newtonian})$.
   $$\rho_p \equiv \rho^* - \rho = \rho(1/\mu - 1) + (4\pi G)^{-1} L\vec{e}_g \cdot \vec{\nabla}|g|.$$
22. The Oort discrepancy: near the sun $\rho^* \approx \rho/\mu(v_\odot^2/ra_\text{o})$.

   a. $\rho^*/\rho \approx \text{const}$.
   b. The local Oort discrepancy factor equals that of the global mass discrepancy at the solar orbit.

23. Negative "dark matter".
24. Light bending and gravitational lensing.

1. Introduction

I want to present to you a solution to the mass discrepancy problem, that is utterly different from the conventional explanation.

   Masses in galaxies and aggregates of galaxies are deduced from the observed velocities and distances in such systems. A relation that rests on Newton's laws of dynamics, between these quantities and the masses, is used.

   The masses of galactic systems derived in this way (the Newtonian dynamical masses) do not agree with the masses one observes directly. The former are, in general, larger (and in many cases much larger) than the latter.

   Newton's laws have proven to be so reliable in describing laboratory and solar system phenomena (when relativistic effects can be neglected), that there is an overwhelming tendency to apply them in the realm of the galaxies as well. The observed mass discrepancy is thus perceived as evidence for the existence of dark matter in galactic systems.

I have adopted the opposite view and taken the mass discrepancy to manifest the breakdown of Newtonian dynamics under the conditions that prevail in galaxies.

Astrophysicists consider it bad form to challenge established laws of physics and, on the whole, have been reluctant to spend much effort on possible alternatives. Nonetheless, it is high time that we seriously study the possibility that the cosmic mass discrepancy reflects the inadequacy of Newtonian dynamics.

What may be the motivation for following such a route? Firstly, the conventional solution — the dark matter hypothesis (hereafter DMH) — leaves much to be desired. The DMH is completely arbitrary in that one invokes the existence of dark matter in just the correct amount and spatial distribution needed to explain the mass discrepancy in each and every case, for itself. The explanation of every phenomena connected with the mass discrepancy requires a separate assumption about the alleged dark matter. It now appears, as we heard earlier, that we need more than one type of dark matter to explain discrepancies on different scales (some say at least three types).

One should also remember that, so far, not a trace of the dark matter has been detected directly (and this is not merely due to the definition of dark matter).

In short, the dark matter hypothesis is very elaborate and yet quite useless. It has not led to better understanding of galaxy dynamics (besides closing the mass gap) and has not helped us find relations between different observed phenomena. These shortcomings of the DMH cannot be taken as evidence against dark matter, but one can, at least, draw some encouragement from them when pursuing iconoclastic alternatives.

There are other facts that justify serious consideration of a breakdown of Newtonian dynamics as the alternative to dark matter.

There now exist high quality data on galaxy dynamics that make it rather easy to test any modified dynamics that is specific enough. Such theories are easily falsifiable by present data. The data exhibit various clear-cut regularities (such as flat rotation curves, the Fisher-Tully and Faber-Jackson relations, etc.) that the DMH leaves unexplained and unrelated.

Secondly, we note that all the mass determinations in galactic systems are based on a *single* dynamical relation. The velocity, v, of a test particle, its distance, r, from the center of an attracting body that it is bound to, and the mass M of that body, which we seek to determine, are related by

$$v^2 = \alpha M G r^{-2}. \tag{1}$$

Here α is a coefficient of order unity which depends on the exact definition of v and r and on the geometry of the system, and G is the gravitational constant. Equation (1) is derived from Newton's second law and from the law of gravity (or the Poisson equation).

It is only needed that this one relation breaks down, under the conditions typical for galactic systems, to invalidate all the mass determination. If the correct modified relation is used to determine the masses, the mass discrepancy will be eliminated and many concrete and unavoidable consequences pertaining to galaxy dynamics will follow.

Of all the assumptions and relations of conventional physics on which the mass determination is based, the first culprit that comes to mind is the distance dependence of the law of gravity at large separations. One may assume that the acceleration, a, of a test particle at a separation, $\vec{r}$, from a (point) mass, M, is given by

$$\vec{a} = -\frac{MG\vec{r}}{r^3} f(r/l), \tag{2}$$

with $f(x) \approx 1$ for $x \ll 1$, so that Newtonian gravity is a good approximation at short distances. For $x > 1$, $f(x) > 1$ and the acceleration produced by an attracting mass is larger than the Newtonian acceleration.

I had considered such a modification in detail, and found it to be in clear conflict with the observations. A few of the main arguments are listed below.

An explanation of the mass discrepancy based on a modified r dependence predicts a mass velocity relation of the form $M \propto v^2$.[5] Such a relation is practically ruled out by the observed Fisher-Tully relation (see below). Also, in order to explain observations in average size galaxies, one must adopt $l \sim 10\,Kpc$. Hence, one expects to find practically no mass discrepancy in galaxies with sizes of a few Kpc or less. Again, this is in conflict with recent observations. In general, one does not find any correlations between the size of a galaxy and the level of mass discrepancy it exhibits (we expect a strong correlation on the basis of Eq. (2)). The observed Oort discrepancy (see below) cannot be explained because it appears on distances much smaller than a Kpc.

2. Dynamics at Low Accelerations

Various other directions had been explored before I proposed a modified law of motion in which the *acceleration* of test bodies is the parameter that determines the degree of departure from Newtonian dynamics.[1]

The basic assumptions of the non-ralativistic version of the proposed dynamics (MOND) may be enunciated as follows: Newtonian dynamics is a good approximation when the accelerations involved are large. In the opposite limit the relation between the acceleration of a test particle at a distance r from a (point) mass M is given by:

$$(a/a_\mathrm{o})\vec{a} \approx -\frac{MG}{r^2}\vec{e}_r \ (\equiv \vec{g}_N). \tag{3}$$

Here $\vec{g}_N$ is the Newtonian gravitational acceleration and a_o as a constant with the dimensions of acceleration, which we must introduce, and $\vec{e}_r$ is a radial unit vector. (Relation (3) replaces the Newtonian low of motion $\vec{a} = \vec{g}_N$.)

The acceleration constant, a_o, is also assumed to play another role, that of the borderline acceleration between the Newtonian and low acceleration regimes. For $a \gg a_\mathrm{o}$, the Newtonian relation is a very good approximation, whereas the asymptotic relation (3) holds for $a \ll a_\mathrm{o}$.

Matters can be complicated by positing a modification whose departure from conventional physics depends on both distance and acceleration or one that involves more than a single new constant (for instance, by adopting a more complicated dependence of the inertia term on acceleration). There are no observations, at present, that require such elaborateness and we should adhere to the more parsimonious as long as possible.

Also, there are various theories that one can build on the assumptions listed above. Fortunately, many major consequences and predictions of MOND follow directly from the basic assumption.[a]

Even before deciding on the exact theory, one encounters the question of the interpretation of MOND. Is it to be interpreted as a modification of the second law so that the inertia force F_i is quadratic in the acceleration, for $a \ll a_o$, instead of the conventional relation $F_i = ma$? Or perhaps it signifies a breakdown of Newtonian gravity, leaving the second law intact.[1]

To make the latter interpretation appear more transparently we write Eq. (3) in the form

$$\vec{a} \approx - \left(\frac{MGa_o}{r^2} \right)^{1/2} \vec{e}_r. \tag{4}$$

Thus, the gravitational force on a test mass m at a distance r from a (point) mass M is given by

$$F(m \ll M, M, r) \approx \begin{cases} m(MGa_o)^{1/2}r^{-1} & MGr^{-2} \ll a_o \\ mMGr^{-2} & MGr^{-2}L \gg a_o \end{cases}. \tag{5}$$

The two interpretations are drastically different of course. The former entails a departure from Newtonian dynamics whenever the acceleration is small, no matter what combination of forces produces it. According to the latter, deviations from Newtonian gravity are expected only when the gravitational acceleration is much smaller than a_o.

We do not know, at present, which interpretation is to prevail. The only existing relevant data describe dynamical behavior of galactic systems where gravity is the only force of importance. Such data do not help us decide between the two interpretations. The theory of MOND that we shall present below is based on a breakdown of Newtonian gravity.

The value of a_o was determined in a few independent ways which I shall describe in the next lecture. All gave a value in the range $(1–8) \times 10^{-8} \mathrm{cm\,s^{-2}}$. (The deduced value of a_o depends on the values one adopts for the Hubble constant, H_o, and for the mass-to-luminosity ratio of the stellar component. Most of the uncertainty in the value of a_o can be imputed to our ignorance of these astronomical parameters.)

---

[a]One then has to assume in addition that a small object of a small mass moving in a field of a large mass can be considered a test particle even if the former is made up of sub-particles that have high (internal) accelerations. Any theory of MOND should satisfy this additional requirement.

This range of values contains that of $cH_o \approx 5 \times 10^{-8}(H_o/50\,\mathrm{km\,s^{-1}\,Mpc^{-1}})\mathrm{cm\,s^{-2}}$, were H_o is the Hubble constant.

This near equality between a_o and cH_o may be of great importance. Firstly, if this relation is valid at all times, a_o must vary on cosmological time scales. This will affect strongly our view of galaxy evolution on such time scales.

Secondly, when we find a constant that appears in the equations of local dynamics (in small systems) and equals to one related to cosmology, we are immediately reminded of the Mach principle.

According to this principle, in local systems, whose size is small compared with the size of the horizon, the dynamics is strongly affected by the interaction of the system's constituents with the content of the rest of the universe. This interaction is not apparent in the laws of physics we use because one is able to write down an effective theory that involves only variables of the small system itself. This theory is only approximate and the interaction with the ambient universe comes in through some of constants in the reduced theory. These constants can be calculated, in a more general theory, from global cosmological properties. (This state of affairs is much like that in a small isolated laboratory on the surface of the earth, where one deduces that all free bodies move with a constant acceleration g, which is then perceived as a constant of nature. Once one is free to leave the confine of the laboratory, he discovers that g is calculable from parameters of the earth, and indeed that it loses its significance as a constant of nature.)

In the theories that will be described below, a_o is put in by hand, and the fact that it nearly equals cH_o is not brought to bear. It is hoped that in a future theory, MOND will result as an effective theory and the value of a_o will be calculable from cosmological parameters.

Obviously, if the Machian point of view expressed above is valid, one may not use, the local effective theory to describe cosmology. One should be cognizant of this fact even when employing a relativistic version of MOND to discuss cosmology. It is for this reason that I have been reluctant to try and apply some variant of MOND to cosmology.

Note, in this connection, that an acceleration, a, defines a scale of length $r_a = c^2/a$. This, for instance, is the transition radius from the near field to the radiation zone of an accelerated charge. Also, an accelerated observer has a region of space-time from which it is causally disconnected. The distance from the observer to the boundary of this region is given by r_a.

Now, when a is much larger than a_o, r_a is much smaller than the distance to the cosmological horizon. If $a \ll a_o$ the horizon is within r_a of the system.

The distance to the cosmological horizon, or alternatively, the expansion rate at a given epoch, defines a quantity with the dimension of an acceleration. At the present epoch this quantity has the value that we find for a_o from studies of galaxy dynamics.

We may indeed expect that in a theory that is based on Mach's principle, the dependence of the inertia force on acceleration for $a \ll a_o$ is different than that at $a \gg a_o$.

MOND is occasionally mistaken for a modification of gravity at large separations. This is far from being the case even if we interpret MOND as a modification of gravity. It is true that for a given attracting mass, M, the acceleration changes from Newtonian to non-Newtonian as the separation increases. However, the transition does not occur at a universal radius (as is the case for modification that is described be Eq. (2)). Rather, the transition occurs at a universal acceleration, a_o. The transition radius $r_t \equiv (MG/a_o)^{1/2}$ depends on the attracting mass. Both the radius and the mass dependence is modified.

Unlike theories that are described by Eq. (2), MOND gravity is a non-linear theory. The acceleration produced by a collection of masses is not the sum of the accelerations due to each mass separately.

3. A Nonrelativistic Formulation

On the basis of the assumptions of MOND we built a nonrelativistic Lagrangian theory (Ref. 2), which I shall now outline. The acceleration field, $\vec{g}(\vec{r})$, of a test particle in the gravitational field of a mass distribution, $\rho(\vec{r})$, is taken to be derivable from a potential, $\varphi(\vec{r})$, i.e. $\vec{g} = -\vec{\nabla}\varphi$. The field equation for φ is a generalization of the Poisson equation $[\nabla^2\varphi = 4\pi G\rho]$:

$$\vec{\nabla} \cdot [\mu(|\nabla\varphi|/a_o)\vec{\nabla}\varphi] = 4\pi G\rho, \tag{6}$$

which we must provide with appropriate boundary conditions. Equation (6) is derivable from the Lagrangian

$$L = -\int d^3r\{\rho\varphi + (8\pi G)^{-1}a_o^2\mathcal{F}[(\nabla\varphi)^2/a_o^2]\}. \tag{7}$$

The function $\mu(x)$ is given by $\mu(x) = [d\mathcal{F}(y)/dy]_{y=x^2}$, and one chooses $\mathcal{F}(y)$ so that

$$\mu(x) \approx \begin{cases} x & x \ll 1 \\ 1 & x \gg 1 \end{cases}.$$

Given a source mass distribution $\rho(\vec{r})$ one solves Eq. (6) to obtain the potential and acceleration fields. The solution of Eq. (6) exists and is unique in any volume, V, within which $\rho(\vec{r})$ is given, and on the boundary of which either φ or $\mu(|\nabla\varphi|/a_o)\partial_n\varphi$ is given.[3] Here $\partial_n\varphi$ is the derivative of φ normal to the boundary.

The motion of a test particle in the field of $\rho(\vec{r})$ is given by the equation of motion $\ddot{\vec{r}} = \vec{g}(\vec{r})$. The force acting on a finite (non-test) body that occupies a sub-volume u of the system is

$$\vec{F}_u = \int_u \rho(\vec{r})\vec{g}(\vec{r})d^3r, \tag{8}$$

and the acceleration of its center of mass, $\vec{R}$, can be shown to be $\ddot{\vec{R}} = \vec{F}_u/m_u$ where $m_u = \int_u \rho d^3r$ is the mass in u.

The total momentum,

$$\vec{P} \equiv \int \rho(\vec{r})\vec{v}(\vec{r})d^3r$$

as well as the angular momentum,

$$\vec{J} \equiv \int \rho(\vec{r})\vec{r} \times \vec{v}d^3r$$

of a closed system are conserved. Also

$$\dot{E} \equiv \dot{E}_K - \dot{L} = 0,$$

where $E_K \equiv \frac{1}{2} \int v^2(\vec{r})d^3r$.

MOND must satisfy an important requirement in order for it to provide the explanation of the mass discrepancy. Consider a sub-system ("star") in a large system ("galaxy") such that the mass of the star is much smaller than that of the galaxy and the size of the star is much smaller than the extent over which the "galaxy's" field varies. It is required that the center of mass acceleration of the "star" in the "galaxy" will be that given by MOND albeit the very large (gravitational) accelerations exerted by elements of the "star" on each other.[1]

This requirement is satisfied by the theory that I have just described.[2] Had this not been the case, different objects such as stars, binaries, HI clouds, etc. would have fallen with different accelerations at the same locations in the galaxy and the weak equivalence principle would have been violated.

I shall now demonstrate that the Lagrangian theory indeed satisfies the assumption of MOND. Eliminating ρ between the MOND equation

$$\vec{\nabla} \cdot [\mu(g/a_\circ)\vec{g}] = -4\pi G\rho$$

and the Poisson equation for the Newtonian fields $\vec{g}N$ and φN (i.e. $\vec{\nabla} \cdot \vec{g}N = -4\pi G\rho$), we get

$$\vec{\nabla} \cdot [\mu(g/a_\circ)\vec{g} - \vec{g}_N] = 0. \tag{9}$$

(We remember, of course, that if MOND is valid then the Newtonian fields $\vec{g}_N$ and φ_N are not the gravitational fields, they serve here only as calculational auxiliaries defined as the solution of the Poisson equation.) The expression in square parenthesis in Eq. (9) is thus a pure curl. In problems of high symmetry such as spherically, cylindrically, or plane symmetric systems such a curl must vanish and we have the exact result:

$$\mu(g/a_\circ)\vec{g} = \vec{g}_N. \tag{10}$$

This relation can also be shown to hold, in general, for the leading power in the inverse distance, at a large distance from a bound mass. Since $\mu(x) \approx x$ for $x \ll 1$ we have in this limit

$$g^2/a_\circ \approx MGr^{-2}, \tag{11}$$

as required (compare with Eq. (3)).

As it turns out, Eq. (10) is a very good approximation, for the acceleration field by test particles, in a wide range of systems (not only those with high symmetry). It may, for instance, be safely used to calculate rotation curves of galaxies.[3] This is very fortunate because Eq. (10) is much easier to solve than the field equation (6). The latter is a nonlinear partial differential equation whereas the former is easily solved once the Newtonian field is known.

4. Effects of an Ambient Field

Now consider the dynamics within a sub-system (for instance, an open cluster) that is freely falling in the ambient acceleration field of a "mother" system (say a galaxy). Assume that the "cluster" is much smaller than the length over which the ambient field varies appreciably, so that tidal effects can be neglected altogether.

In Newtonian dynamics, indeed in any theory that satisfies the strong equivalence principle, the internal dynamics (involving motion in the subsystem relative to its center of mass) are oblivious to the presence of the constant ambient acceleration field. This is not the case in MOND which does not satisfy the strong equivalence principle even in its nonrelativistic form.[1]

Let $\rho(\vec{r})$ be a mass distribution that falls in a constant external acceleration field $\vec{g}_o$. We seek the solution, $\vec{g}$, of the field equation (6) with the boundary condition

$$\vec{g} \xrightarrow{r \to \infty} \vec{g}_o.$$

The center of mass itself falls with acceleration $\vec{g}_o$ and so the internal dynamics are governed by the field $\vec{q} \equiv \vec{g} - \vec{g}_o$.

The fact is that, in general, $\vec{q}$ is very different from the solution of the field equation with the condition

$$\vec{g} \xrightarrow{r \to \infty} 0$$

($\vec{q}$ also vanishes at infinity).

So, an isolated system behaves very differently than it would have in an external field. We demonstrate this point with an example of wide applicability.[3]

Suppose that $|q| \ll g_o$ everywhere, and let $\vec{g}_o$ be in the z direction. We can then linearize the field equation by retaining only the lowest order terms in $\vec{q}$. We do this by writing

$$|\vec{g}_o + \vec{q}| \approx g_o + q_z,$$

$$\mu(|g_o + q|/a_o) \approx \mu(g_o/a_o) + \mu'(g_o/a_o)q_z/a_o.$$

Thus,

$$\vec{\nabla} \cdot \{\mu(g/a_o)\vec{g}\} \approx \mu(g_o/a_o)\vec{\nabla} \cdot \vec{q} + \mu'(g_o/a_o)g_o q_{z,z}/a_o = \mu_o(\vec{\nabla} \cdot \vec{q} + L_o q_{z,z})$$

where $\mu_o = \mu(g_o/a_o)$ and $L_o = (d\ln\mu(x)/d\ln x)_{x=g_o/a_o}$. We find that with $\vec{q} = -\vec{\nabla}\eta$, the intrinsic potential, η, satisfies

$$\frac{\partial^2 \eta}{\partial x^2} + \frac{\partial^2 \eta}{\partial y^2} + \frac{\partial^2 \eta}{\partial z'^2} = 4\pi \mu_o^{-1} G\rho, \qquad (12)$$

with $z' \equiv z/(1+L_o)^{1/2}$.

Equation (12) is just the Poisson equation for the density $\hat{\rho}(\vec{r}) \equiv \mu_o^{-1}\rho(\vec{r})$ and the coordinates

$$\vec{r}' = (x', y', z') = [x, y, z(1+L_o)^{-1/2}].$$

The effective density $\hat{\rho}$ can be much larger than ρ since $\mu_o^{-1} \approx a_o/g_o$ for $g_o \ll a_o$. Hence, the internal field around a mass that produces a small perturbation on an ambient field is quasi-Newtonian (i.e. it decreases like r^{-2} at large distances). The field corresponds to a mass larger than the one actually present by a factor $1/\mu(g_o/a_o)$, and is aspherical.

For example, far away from a point mass (e.g. the sun) that is falling in an external field (e.g. that of the galaxy), the acceleration of test particles (comets) relative to the sun is described by an effective field with elliptical equi-potential surfaces $[x^2+y^2+z^2/(1+L_o) = $ const] elongated in the direction of the galactic field. The field of the sun is thus neither spherical nor radial. This asphericity becomes most prominent at distances from the sun larger than r_s, where r_s is given by

$$\mu(g_o/a_o)g_o = M_\odot G r_s^{-2},$$

and g_o is the acceleration of the sun in the galaxy.

When the external acceleration is much larger than a_o we have $\mu_o \approx 1$, and $L_o \ll 1$ [because $\mu(x \gg 1) \approx 1$]. The resulting internal dynamics is very nearly Newtonian in this case. For example, an experiment in a terrestrial laboratory involving small relative accelerations will show practically no MOND effects because it is immersed in the field of the earth ($g > 10^{10}a_o$).

In summary, if a subsystem with a typical internal acceleration, g_{in}, falls in an external field with an acceleration, g_{ex}, the internal dynamics in the subsystem depends on the relative magnitudes of g_{in}, g_{ex}, and a_o. When the inequalities between every two of these quantities are strong, and the subsystem is small compared with the distance over which g_{ex} varies substantially, the description of the internal dynamics takes a relatively simple form.

When either $g_{in} \gg a_o$ or $g_{ex} \gg a_o$ the internal dynamics is Newtonian. When $g_{in} \gg g_{ex}$ the system is isolated and MOND dynamics hold. When $g_{in} \ll g_{ex} \ll a_o$ the internal dynamics is quasi-Newtonian as we demonstrated above. Figure 1 shows, schematically, where various astronomical systems fall in the $g_{in} - g_{ex}$ plane.

More details of the properties of the field equation and its solutions can be found in Refs. 2 and 3.

We still lack a satisfactory relativistic theory with the MOND non-relativistic behavior. If a massive system of size R is to be both relativistic and of low acceleration, R should satisfy $R > c^2/a_o$. Because of the near equality $a_o \sim cH_o$ this

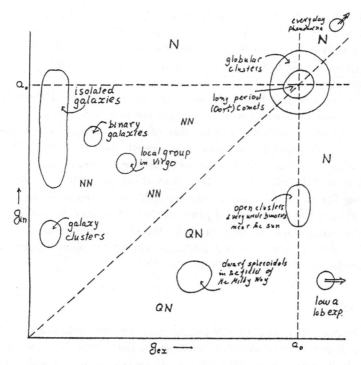

Fig. 1. A classification of various systems according to their intrinsic and center-of-mass accelerations.

implies: $R > cH_o^{-1}$ (the cosmological horizon's size). We are then dealing with cosmology. We also find that a charge that is accelerated at a rate smaller than a_o, has it's radiation zone beyond the cosmological horizon because then,

$$c^2/a > c^2/a_o \sim cH_o^{-1}.$$

We also need a relativistic extension in order to describe light bending and gravitational lensing which I discuss below.

5. Observational Consequences

Employing MOND, we find that a given mass distribution produces, on average, larger accelerations than those dictated by Newtonian theory. Hence, if we insist on using Newtonian dynamics, we will have to assume larger masses to explain the observed accelerations. This, according to MOND, is the origin of the mass discrepancy.

I believe that the key to the mass discrepancy problem lies in observing and understanding galaxies, notably disc galaxies, in spite of the fact that other systems, such as galaxy cluster, show wider discrepancies. The former are more regular in shape, can be observed and interpreted more easily and with higher confidence. This is especially true for discs where the motion of the test articles (neutral hydrogen, HII regions, etc.) can be shown to move in very nearly circular orbits.

In addition, observed properties of galaxies exhibit very clear-cut regularities that are of great significance in providing strong constraints on any explanation of the mass discrepancy.

I shall thus concentrate here only on some consequences of MOND that pertain to galaxies. Before doing so I just wish to state the results for binary galaxies, small groups and clusters of galaxies, and for the infall of the local group into Virgo. A reanalysis of the observed dynamics of these systems in light of MOND,[4] using the same value of a_o as obtained from the analysis of galaxies, eliminates the need to invoke dark matter. In other words, the observed masses suffice to hold the systems from breaking apart. All the aforementioned galactic systems produce acclerations $(a \ll a_o)$ and thus the exact form of $\mu(x)$ or, for that matter, the exact form of the theory we use is immaterial for their analysis. We are working in the deep asymptotic regime and deal with motion of test particles (except for binary galaxies) and can thus use directly the basic assumptions of MOND.

5.1. *Disc Galaxies*

Rotation curves: Rotation curves constitute a major prediction of MOND, one that can be tested in most detail with present capabilities. Since we assume that there is no dark matter in appreciable quantities, we should be able to derive the mass distribution from the observed light distribution, assuming a constant M/L value for each observed component of the galaxy (disc or bulge). (I am using the term "light distribution" loosely. The mass distribution can have contributions from neutral hydrogen emitting a 21 cm line, hot gas emitting X-rays, etc.) Given the mass distribution, we calculate the rotation curve from MOND. This calculated curve should agree with the one observed. The failure of such a comparison when we use Newtonian dynamics is one manifestation of the mass discrepancy problem.

Some rotation curves calculated from MOND for model galaxies are shown in Fig. 2 (taken from Ref. 5). The MOND velocity curve calculated from the light distribution of NGC 3198 (M/L of the disc is the only free parameter) is shown in Fig. 3 together with the data points (taken from Ref. 6).

A complete rotation curve test on a galaxy is sometimes too demanding a task because it requires a reliable photometry (if possible in two colors to check on variations in M/L), a good rotation curve, knowledge of the gas mass distribution etc. There are consequences of MOND that can be tested on galaxies for which only partial information is available. I give here two examples:

(1) If $v(r)$ is the circular velocity around a bound mass M in an orbit of large radius r, the acceleration $v^2(r)/r$ is given according to the assumptions of MOND[2] by

$$(v^2/r)^2/a_o \approx MGr^{-2}, \tag{13}$$

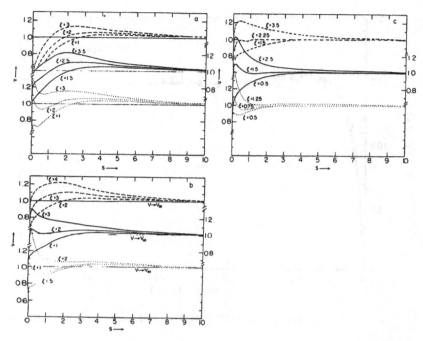

Fig. 2. Rotation curves for model galaxies calculated with Eq. (10) (see Ref. 5 for details). The model galaxies are made of an exponential disc and a spherical halo. Each group of three curves belongs to galaxies with the same mass distribution but different surface densities. Groups differ by the ratio of disc to bulge mass and size.

or

$$v^4 \approx MGa_o \equiv v_\infty^4(M). \tag{14}$$

Hence, the velocity approaches a constant value far outside the attracting mass. The rotation curves of all *isolated galaxies* are thus flat asymptotically. The isolation criterion requires that, at r, the acceleration due to the galaxy dominates any external acceleration (e.g. the one the galaxy falls with, in a group or a cluster). The rotation velocity will start to decline around the radius r_i where the external acceleration is comparable with the internal one.

The asymptotic flatness of galaxy rotation curves has been built into the basic assumption of MOND (by requiring that the inertia term be quadratic in a for small a). One may then claim that the verification of flatness at larger and larger galaxy radii, and for more and more galaxies, should not be deemed a success of MOND. However, the elevation of asymptotic flatness from an observed property of a number of galaxies known at the time MOND was posited (at moderate radii), to a general property of all isolated galaxies, is cornerstone of MOND. Observed violation of this assumption will certainly rule out MOND. Its verification should be taken as a strong support of the breakdown of Newtonian

M. Milgrom

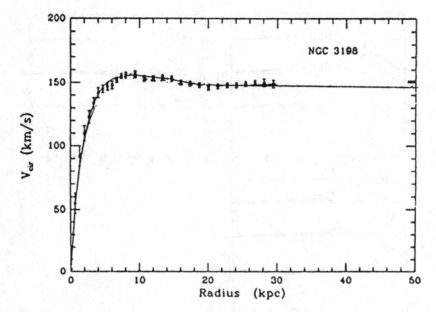

Fig. 3.   The MOND rotation curve for NGC 3198 (line) and the data points (from Ref. 6).

dynamics as an explanation of the mass discrepancy, and of a basic (albiet implicit) assumption of MOND.

(2) We learn from Eq. (14) that the asymptotic circular velocity depends only on the total mass of the galaxy. The total mass can be obtained from the total "luminosity" (no need for a detailed surface photometry) up to the uncertainties in the stellar M/L, etc.

Take a sample of galaxies that the luminosity and the plateau velocity are known for, and such that M/L does not vary much over the sample (at least not systematically with L). We can then plot $\log L$ versus $\log v$ (the Fisher-Tully relation). MOND predicts a slope of exactly 4.

Here we encounter the first method of determining the value of a_o from the intercept of the $\log L$-$\log v$ relation. Assuming some theoretical M/L values for galaxies we obtained $a_o \approx 2 \times 10^{-8}(H_o/50\,\mathrm{km\,s^{-1}M\,pc^{-1}})^2\mathrm{cm\,s^{-2}}$.[5] This value scales like inverse of the (M/L) value we adopt.

Surface densities: The constant a_o, which is introduced by MOND, defines a value of surface density $\Sigma_o \equiv a_o G^{-1}$. (For $a_o = 5 \times 10^{-8}\,\mathrm{cm\,s^{-2}}$ we have $\Sigma_o \equiv 0.7\,\mathrm{g\,cm^{-2}}$.)

This value of the surface density plays a special role in galaxy dynamics. We shall see later that it is roughly the maximal average surface density that an isothermal sphere can have. In discs, the average surface density cannot exceed Σ_o appreciably without the rotation curve acquiring a large hump (see Fig. 2 in which $\Sigma \gtrsim \Sigma_o$

correspond to $\xi \gtrsim 1$). Such humps are not observed, and so MOND implies that the distribution of average surface densities of disc galaxies should be cut off roughly above Σ_o.

Comparing this value with the observed cut-off that is indeed observed (the Freeman law) we get a second and independent determination of a_o[5] $a_o \sim 2 \times 10^{-8}\,\mathrm{cm\,s^{-2}}$. (The resulting value of a_o scales like the theoretical values of M/L we adopt for galaxies.)

In this connection, note that in Newtonian dynamics the shape of the rotation curve of a galaxy (presence of hump, etc.) depends only on the *shape* of the mass distribution but not on its normalization (as expressed, say, by the value of the average surface density).

In MOND, the shape of the rotation curve depends on the total mass too. This is because the amount of mass the galaxy contains within a certain radius determines the acceleration which, in turn, determines how strong the departure from Newtonian dynamics is. We can see this clearly in Fig. 2 where rotation curves are given for sets of galaxy models with the same mass distribution but different surface densities, resulting in different rotation curves.

Another important prediction of MOND concerning "dark matter" inside galaxy discs, will be discussed below after we develop some more tools.

5.2. *Elliptical Galaxies*

Our hypothesis is that galaxies contain no appreciable quantities of dark matter. We should thus, in principle, be able to understand the observed light distribution and velocity dispersions in a consistent manner.

However, as Scott Tremaine explained to us, this is a practically impossible task unless we can make some strong assumptions about the distribution of stellar orbits in the galaxy. Such assumptions substitute our knowledge that the trajectories in disc galaxies are circular.

We may, for instance, approximate elliptical galaxies (and, for that matter, galactic bulges or cluster of galaxies) by isothermal spheres (hereafter IS). The rationale behind such an approximation is even more sound in the case of MOND than in Newtonian dynamics. According to the latter, isothermal spheres are necessarily infinite in mass, so one must assume some artificial cutoff device in order to make them into acceptable models.

All MOND IS have a finite mass[7] (and their density always decreases like a power law at large radii $\rho \propto r^{-\alpha}$ with $\alpha > 3$). The reason for this is easy to see. In Newtonian dynamics, the escape speed from any point in the field of a finite mass is finite. In an IS there are, at every point, particles with arbitrarily high velocities that would have escaped from the system had it been of finite mass.

In MOND the escape speed is everywhere infinite. (To see this note that from the basic assumptions of MOND the gravitational potential far from a finite mass increases logarithmically: $\varphi(r) \to v_\infty^2 \ln(r)$, where v_∞ is the asymptotic circular velocity around that mass). Thus, particles at arbitrarily high velocities do not escape and IS are of finite mass.

What other properties do MOND IS have?

We find that there is a maximum average surface density that an IS can have[7] [the surface density defined, say, as the total mass divided by the area containing half the total mass projected on the sky]. The upper limit is the critical surface density we met before: $\Sigma_o = a_o G^{-1}$.

Again, the role of Σ_o as limiting density can be deduced from the basic assumptions of MOND as follows: Let $r_t = (MG/a_o)^{1/2}$ be the transition radius of an isothermal sphere of mass M. The statement we made above amounts to the assertion that the radius, $r_{1/2}$, containing half the projected mass cannot be much smaller than r_t. If most of the mass is contained within r_t, we get an IS that is Newtonian (because, by the definition of r_t the acceleration, in most of the sphere, is larger than a_o). Such spheres are infinite, and so $r_{1/2}$ cannot be substantially smaller than r_t.

The cut-off found in the distribution of surface brightnesses of ellipticals is the observation pertinent to this property of IS.

The total mass and velocity dispersion (temperature) of MOND IS obey a relation analogous to the $M - v_\infty$ relation for discs. We studied IS with constant radial dispersion, σ_r, and tangential dispersion, σ_t.[7] With $\beta \equiv 1 - \sigma_t^2/\sigma_r^2$, the three dimensional dispersion is

$$\sigma^2 \equiv \sigma_r^2 + 2\sigma_t^2 = (3 - 2\beta)\sigma_r^2.$$

We find that

$$a_o GM = Q\sigma^4, \tag{15}$$

where Q is not a constant but varies, within a limited range $1 \leq Q \leq \frac{9}{4}$, among the isothermal spheres. The upper end of this range is obtained for low acceleration (or low surface density) IS with $g \ll a_o$ everywhere in the sphere (or for which $r_{1/2} \gg r_t$). For those, $Q \approx \frac{9}{4}$ and is independent of β. If galaxy clusters can be modelled by IS they would fall in this category.

The relevant observational phenomenon is the Faber-Jackson relation between the luminosity of an elliptical and its line-of-sight central velocity dispersion. When comparing Eq. (15) with observations one should remember that σ is the 3-dimensional dispersion and not the average dispersion observed along a given line of sight (the two are proportional to each other when $\beta = 0$). More details pertinent to the $M - \sigma$ relation are to be found in Ref. 7.

Another approach that may be taken, in studying ellipticals, is based on the use of test particles (other than stars) falling in the fields of these galaxies. One can, for instance, measure the density and temperature distribution of an X-ray emitting envelope around ellipticals. The collisional mean-free-path in the gas is short and hence the velocity distribution may be taken as Boltzmanian and isotropic. The distribution of pressure and density in the gas gives us the gravitational field if the gas is in hydrostatic equilibrium. The measured field can then be compared with the prediction of MOND as deduced from the distribution of the observed mass. Preliminary observational results are described in Ref. 8 and comparison with MOND in Ref. 9.

Some ellipticals may possess a gas disc in which rotational velocities can be measured (e.g. Ref. 10). Studies of shells around ellipticals may also be used to extract information on their potential fields.[11]

5.3. *The General Expression for the "Dark Matter" Density*

Let me now derive the general relation between the actual mass distribution and that of the dark matter (or more appropriately, the phantom matter).

If MOND is correct, the acceleration field $\vec{g}(\vec{r})$ that will be measured about a true mass distribution $\rho(\vec{r})$ will be related to ρ by

$$\vec{\nabla} \cdot [\mu(g/a_o)\vec{g}] = -4\pi G\rho, \tag{16}$$

or equivalently by

$$\mu(g/a_o)\vec{\nabla} \cdot \vec{g} + a_o^{-1}\mu'(g/a_o)\vec{g} \cdot \vec{\nabla}g = -4\pi G\rho, \tag{17}$$

where $g \equiv |\vec{g}|$. Note that this relation holds for both of the MOND formulations Eqs. (6) and (10) given above.

One measures $\vec{g}(\vec{r})$ and, adhering to Newtonian laws, deduces the mass distribution $\rho^*(\vec{r})$ via the Poisson equation:

$$\begin{aligned}\rho^* &= -(4\pi G)^{-1}\vec{\nabla} \cdot \vec{g} = \rho/\mu(g/a_o) + (4\pi G)^{-1}a_o^{-1}(\mu'/\mu)\vec{g} \cdot \vec{\nabla}g \\ &= \rho/\mu(g/a_o) + (4\pi G)^{-1}L(g/a_o)\vec{e}_g \cdot \vec{\nabla}g.\end{aligned} \tag{18}$$

Here $L(x) = d\ln\mu(x)/d\ln x$ and $\vec{e}_g$ is a unit vector in the direction of $\vec{g}$. The density distribution of the phantom matter will thus be

$$\rho_p \equiv \rho^* - \rho = \rho(1/\mu - 1) + (4\pi G)^{-1}L\vec{e}_g \cdot \vec{\nabla}g. \tag{19}$$

The phantom matter will, in general, be found to be distributed differently from the "luminous" (actual) matter. This is because $\mu(g/a_o)$ varies from point to point and because of the presence of the second term in Eq. (19).

A general question comes to mind in this context. Given a mass distribution $\rho(\vec{r})$ (or the energy momentum tensor in a relativistic problem) can one always add a fictitious dark matter (energy-momentum density) so that the results of MOND are obtained with Newtonian dynamics (general relativity). In other words, can an observed mass discrepancy always be explained by dark matter even if the true explanation is the departure entailed by MOND?

The answer to this question is negative. One possible counterexample involves light bending and will be discussed later, another one follows.

5.4. *The Sign of the Phantom Density — Negative "Dark Matter"*

There are mass configurations for which $\rho_p < 0$ in some regions of space. If a clear-cut instance of this type is found it will rule out the conventional dynamics. This is because the adherence to Newtonian dynamics will entail the acceptance of negative "dark matter", an utterly nonsensical alternative.

Examples of such configurations are given in Ref. 12, and I describe one here. Consider a binary of galaxies, say of equal masses, situated on the z axis with the origin at their midpoint. Starting at the origin where $g = 0$ we go away from it in the x–y plane. At first g increases, and reaches a maximum at a distance r_m from the origin, then it declines. Everywhere in the x–y plane $\vec{g}$ is in the direction of the origin. Thus, $\vec{e}_g \cdot \vec{\nabla}g$ is negative in the x–y plane below r_m. Since $\rho = 0$ in this plane, ρ^* is negative in this region [see Eq. (18)]. A schematic depiction of the whole $\rho^* < 0$ regime, in such a system, is shown in Fig. 4.

5.5. *Phantom Matter in Galactic Discs*

Oort described a technique for measuring the distribution of the dynamical mass, in the direction perpendicular to the disc of the Milky Way. John Bahcall gives a detailed description of the method in his lectures, as well as the results of his own improved analysis (see also Ref. 13).

I shall now give the relation that one expects on the basis of MOND, between the actual and the Newtonian dynamical density distributions of this system. Consider a disc galaxy observed near the galactic plane. At a distance r from the galactic center g is of order $\mu^{-1}M(r)Gr^{-2}$, and so the second term in Eq. (18) is of the order of $\hat{\rho}/\mu$ where $\hat{\rho}$ is the average density of the galaxy within radius r. In the region of the disc where the average density of the galaxy is negligible compared with the local disc density we have from Eq. (18):

$$\rho^* \approx \rho/\mu(g/a_{\rm o}). \tag{20}$$

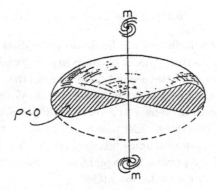

Fig. 4.   A schematic view of the negative phantom density region in a system of two (equal mass) galaxies.

This condition holds within a few hundred parsecs from the galactic plane, in the solar neighborhood. Also, g hardly varies in the direction perpendicular to the disc's plane and equals $v^2(r)/r$ where $v(r)$ is the rotational velocity in the disc. We thus predict the following results of an analysis of the Oort problem near the sun.

(a) The Newtonian dynamical mass is larger than the actual mass by a factor $1/\mu(v^2/ra_o)$.
(b) The phantom density is distributed *in the same way as the actual mass* (up to a height, above the galactic mid-plane, where the second term in Eq. (18) becomes important). This means that the discrepancies in the central density and the surface densities are equal.
(c) The factor $1/\mu(v^2/ra_o)$ is also very nearly the ratio between the observed rotation velocity squared and that deduced from the observed luminous mass in Newtonian dynamics.

Prediction (c) connects two seemingly unrelated quantities (the local disc discrepancy and the global galactic discrepancy within the solar orbit) in a way that is independent on the value of a_o or the form of $\mu(x)$. A similar discrepancy factor is predicted for low surface density open clusters in the solar neighborhood.

The results of the analysis described by Bahcall can be used to determine a_o in yet another way, by requiring that the observed discrepancy factor equals that given in (c) above. For a discrepancy factor of $0.5 \lesssim \rho_p/\rho \lesssim 1.5$, as found by Bahcall, and accounting for the uncertainties in v and the form of $\mu(x)$, we obtain $a_o \approx (1-8) \times 10^{-8}\,\mathrm{cm\,s^{-2}}$.

Lastly, I want to discuss light bending and gravitational lensing. Gravitational lensing of quasar images by galaxies and clusters involves light rays that go through the outskirts of galaxies and are bent in their field. The relevant accelerations are small and we thus ask how lensing is affected by MOND.

Wanting a relativistic theory of MOND, we cannot answer this question. However, the phenomenon of lensing holds a promise to provide a crucial test of MOND, the verification of which may conflict with the theory of general relativity. If ρ^* is the (fictitious) density distribution needed to explain the trajectories of massive particles in the field of a true distribution ρ (as deduced e.g. from a rotation curve), it is likely that a different ρ^* will be needed to explain the trajectories of massless particles. If this is found to be the case, the dark matter hypothesis and indeed the conventional dynamics will be ruled out (more details in Ref. 1).

To summarize, there are many consequences of MOND that are amenable to direct observations. Many of these results can be mimicked with dark matter. Some, however, are inconsistent with the conventional dynamics even if one allows for the presence of dark matter.

References

Milgrom, M. (1983) "A modification of the Newtonian dynamics as a possible alternative to the hidden mass hypothesis", *Astrophys. J.*, **270**, 365.

Bekenstein J. and Milgrom, M. (1984) "Does the missing mass problem signal the break-down of Newtonian gravity"?, *Astrophys. J.*, **286**, 7.

Milgrom, M. (1986) "Solutions for the modified Newtonian dynamics field equation", *Astrophys. J.*, **302**, 617.

Milgrom, M. (1983) "A modification of the Newtonian dynamics: Implications for galaxy systems", *Astrophys. J.*, **270**, 384.

Milgrom, M. (1983), "A modification of the Newtonian dynamics: Implications for galaxies", *Astrophys. J.*, **270**, 371.

van Albada, S., Bahcall, J. N., Begeman, K., and Sancisi, R. (1985) "Distribution of dark matter in the spiral galaxy NGC 3198", *Astrophys. J.*, **295**, 305.

Milgrom, M. (1984) "Isothermal spheres in the modified dynamics", *Astrophys. J.*, **287**, 571.

Forman, W. Jones, C., and Tucker, W. (1985) "Hot coronae around early-type galxies", *Astrophys. J.*, **293**, 102.

Milgrom M. and Bekenstein, J. (1987) "The modified Newtonian dynamics as an alternative to hidden matter," in *Proc. IAU Symp. #117 on Dark Matter in the Universe*, eds. J. Kormendy and J. Knapp.

Lake, G., Schommer, R. A., and van Gorkom, J. H. (1987) "The HI distribution and kinematics in four low-luminosity elliptical galaxies", *Astrophys. J.*, **314**, 57.

Hernquist L. and Quinn, P. J. (1987) "Shells and dark matter in elliptical galaxies", *Astrophys. J.*, **312**, 1.

Milgrom, M. (1986) "Can the hidden mass be negative"?, *Astrophys. J.*, **306**, 9.

Bahcall, J. N. (1984) "Self-consistent determination of the total amount of matter near the sun", *Astrophys. J.*, **276**, 169.

Chapter 12

DARK MATTER IN COSMOLOGY

Anthony Aguirre

Department of Physics,
University of California at Santa Cruz,
Santa Cruz, CA 95064, USA

The last two decades in cosmological research have been an exciting time, and produced an exciting product: we now have in hand a "standard model" of cosmology. While several aspects of this model remain mysterious, its predictions are in remarkable accord with a vast range of observational data. A key aspect of this model, and one of the aforementioned mysteries, is the dark matter: a cold, collisionless consituent of the universe with $\sim 30\%$ of the cosmic energy density. In this article I broadly review the standard cosmological model, and the role and place of (non-baryonic) dark matter in it.

1. Introduction

Since the lectures in this volume were given, there has been great progress in our understanding of the role of, evidence for, and constraints on, dark matter. While we still have no real idea what dark matter is (and indeed must now postulate a new "dark energy" component of unknown nature as well), a rather precise and increasingly well-tested (and testable) picture of dark matter's role in cosmology has emerged.

While our understanding of all of the issues discussed in the Jerusalem Winter School lectures has been advanced, I will focus on the topic which has perhaps advanced the most, and received least attention in the original lectures: the role of cold dark matter (CDM) in the formation of large-scale structure and galaxies. This is a vast subject and here I hope only to give an overview of the "big picture" and indicate directions for further study. Likewise, I have made no attempt to make comprehensive references; I have instead given for most subjects a few references that I find particularly seminal or useful.

I will first review in Sec. 2 the initial conditions for the standard cosmological model, and outline our theoretical understanding of the role of dark matter in structure formation. I will then discuss the confrontation of this theory with observations of the Cosmic Microwave Background (CMB) (Sec. 3), the Lyα forest and the large-scale distribution of galaxies (Sec. 4). Finally I will address the general picture of galaxy formation in CDM cosmology in Sec. 5.7.

2. Dark Matter and Structure Formation

2.1. *Initial Conditions and the Standard Cosmological Model*

The current standard model of cosmology posits that at a very early time, the universe was nearly homogeneous and isotropic, radiation-dominated, and nearly flat. Its geometry is thus described by the Friedmann-Robertson-Walker (FRW) metric

$$ds^2 = dt^2 - a^2(t)[dr^2 + r^2(d\theta^2 + \sin^2\theta d\phi^2)], \tag{1}$$

where $a(t)$ is a scale factor evolving according to

$$\frac{\ddot{a}}{a} = -\frac{4\pi G}{3}(\bar{\rho} + \bar{p}) \tag{2}$$

in terms of averages of the density ρ and pressure p. Two galaxies at small fixed *comoving* separation Δr will have physical separation $d = a(t)\Delta r$, and move apart physically at a rate $v = \dot{a}\Delta r = Hd$, where $H = \dot{a}/a$ is Hubble's constant. The observation of this relation led, of course, to the development of the big-bang cosmology.

Deviations from homogeneity are described by a random variable $\delta(\vec{x}, t)$, defined as

$$\delta(\vec{x}, t) \equiv \frac{\rho(\vec{x}, t) - \bar{\rho}}{\bar{\rho}}, \tag{3}$$

where $\vec{x}$ are *comoving* coordinates (like r, θ, ϕ in the metric) which are fixed for a particle at rest with respect to the cosmic fluid.

These perturbations are generally assumed to be Gaussian, i.e. the Fourier modes

$$\delta_{\vec{k}}(t) \equiv \int d^3\vec{x}\exp(i\vec{k}.\vec{x})\delta(\vec{x}, t) \tag{4}$$

at fixed t are each described by a Gaussian probability distribution of zero mean and variance σ_k (note that k is a *comoving* wavenumber and has units of inverse length). When computing statistical properties of δ for large volumes we can approximate[29] $\sigma_k^2 \simeq |\delta_k|^2$; the latter is often referred to as the power spectrum and taken to have a power-law form:

$$P(k) \equiv |\delta_k|^2 = Ak^n. \tag{5}$$

Such a power spectrum can be translated into a more physically suggestive measure by integrating $|\delta_k|^2$ for modes below the inverse of some length scale r; then the variation in mass M within a sphere of radius r goes as

$$\frac{\Delta M}{M} \propto M^{-(n+3)/6}. \tag{6}$$

Theoretically, n could take a number of values: $n = 4$ is maximal in that the non-linear but momentum-conserving dynamics of particles on small scales would build a large-scale $n = 4$ "tail" to the power spectrum; $n = 2$ would result from randomly throwing particles down within equal-mass cells, while $n = 0$ (Poisson fluctuations) would correspond to just randomly throwing particles down (see, e.g., Peacock[30]). The *observed* value for small k is $n \simeq 1$, which is called "scale-invariant" because the perturbation to the Newtonian gravitational potential is equal on all length-scales; it also has the pleasing property that the perturbation amplitude on the scale of the cosmological horizon is always the same.[31]

This described flat, homogeneous universe with $n = 1$ Gaussian density perturbations is widely thought to have resulted from a period of inflation in the early universe (see the article by Press and Spergel in this volume), though in principle some other process could give rise to it — the reader is encouraged to look for one!

The remainder of the cosmological model is then specified by describing the material and energetic contents of the cosmic fluid at some early time. This is conveniently and conventionally done in terms of the ratio Ω_i of the ith species' present day energy density to the current critical energy density $\rho_{\text{crit},0} \equiv 8\pi G/3H_0^2$, where H_0 is the current Hubble constant. Extrapolation of each density component to a smaller $a(t)$ (e.g., $\rho \propto a^{-3}$ for pressureless matter) then gives each energy density at earlier times.

Current observations (to be described below) indicate that our universe contains $\Omega_r \sim 10^{-5}$ in radiation, $\Omega_b \simeq 0.04$ in baryons, $\Omega_{\text{dm}} \simeq 0.23$ in cold, collisionless, non-baryonic particles (i.e. the Dark Matter), and $\Omega_{\text{DE}} \simeq 0.73$ in some yet-more-enigmatic substance called "Dark Energy" with $p \simeq -\rho$. The repeated postulation of mysterious substances is tolerated by most cosmologists only because of the striking success of the theory these postulates engender.

2.2. *Evolution of Perturbations*

Understanding the growth of the perturbations δ_k rigorously is an intricate subject requiring a careful treatment of perturbation theory in General Relativity; see Padmanabhan[29] for a detailed treatment. It can, however, be understood at two less rigorous but more tractable levels.

The first is somewhat heuristic (though in fact it can be made relatively precise).[31] Consider a density perturbation δ of comoving scale λ in a matter component (i.e. baryonic or dark matter) during a time when the universe can be considered to be dominated either by radiation or by pressureless matter. This can be thought to describe an over/underdense sphere of radius $\sim \lambda$ embedded in a uniform FRW-universe of density $\bar{\rho}$. Birkhoff's Theorem (the relativistic generalization of Newton's "spherical shell" theorem) indicates that the embedding space can be ignored and the sphere treated as an independent universe.[a] If $\delta > 0$ its expansion will be slow relative to the outer region so that its density relative to $\bar{\rho}$ (i.e. δ) will

---

[a] Actually, it states only that a spherically symmetric vacuum solution to Einstein's equations is the Schwarzschild solution; but it can safely be interpreted in this more liberal way.

increase if the inner and outer regions are compared at a later time in such a way that the expansion rates are equal.[b] Working this out reveals that $\delta \propto a^2$ during radiation domination ($a \propto t^{1/2}$), and $\delta \propto a$ during matter domination ($a \propto t^{2/3}$). If λ exceeds the horizon length λ, this analysis captures much of the dynamics, since on these scales different fluid components cannot evolve separately, and pressure support cannot prevent the growth of perturbations (instead, pressure adds to the source term for Einstein's equations).

In either a matter- or radiation-dominated epoch, the horizon grows as t and hence faster than a, so any perturbation of fixed comoving scale will eventually enter the horizon if the epoch last sufficiently long. At this point two new effects become important. First, pressure: prior to horizon entry the sound crossing time across a perturbation ($> \lambda/c$) always exceeded the dynamical time $\sim t$; now it will not if

$$\lambda < \lambda_J \equiv \sqrt{\pi} \frac{c_s}{(G\rho)^{1/2}}, \tag{7}$$

where c_s is the sound speed (or velocity dispersion, in the case of collisionless particles) of the medium. This leads to a minimal "Jeans mass", dependent upon the temperature and density of the medium, below which fluctuations cannot grow. Second, perturbations in different fluid components — such as matter and radiation, or collisionless and collisional matter — may grow at different rates. For example, dark matter, which interacts negligibly with radiation and baryonic matter, can — and in some situations do — grow even if baryonic perturbations are supported against collapse by their pressure. However, even a pressureless perturbation cannot grow if its dynamical timescale $t_{\rm dyn} = 1/\sqrt{G\rho}$ is longer than the expansion timescale t (which is the dynamical time of the dominant fluid component).

Perturbations inside the horizon can be treated using the equations of motion for a fluid in an expanding universe (for those versed in General Relativity, these can be derived directly from the covariant conservation of the energy-momentum tensor: $\nabla_\mu T^{\mu\nu} = 0$) along with the weak-field version of Einstein's equations:

$$\nabla^2 \Phi = 4\pi G(\rho + 3p/c^2), \tag{8}$$

where $\nabla \Phi$ gives the acceleration of a slowly-moving test particle. In a medium dominated by a non-relativistic fluid with sound speed $c_s \equiv \partial p/\partial \rho$, the analysis

---

[b]Perturbation theory in general relativity can be tricky because of the "gauge" ambiguity in choosing a surface of constant time; e.g. for small perturbations one can always choose a surface in which the universe is homogeneous (see Press and Vishniac[33] for an amusing presentation of some of these issues.) This difficulty can be overcome by carefully choosing a fixed gauge[33] or by working in carefully chosen gauge-independent variables.[2,18]

gives:

$$\ddot{\delta}_k + 2\frac{\dot{a}}{a}\dot{\delta}_k = \delta_k \left(4\pi G\bar{\rho} - \frac{c_s^2 k^2}{a^2} \right) . \tag{9}$$

The appearance of the Jeans length can be seen in the r.h.s.: if the physical wave-length of the perturbation, a/k, does not exceed λ_J, the solution is oscillatory; otherwise the term in c_s can be neglected and using the fact that $\bar{\rho} \propto t^{-2}$, the solution splits into a growing mode $\delta_k \propto t^{2/3}$ and a decaying mode $\delta_k \propto t^{-1}$. For a radiation-dominated phase with pressure gradients neglected,

$$\ddot{\delta}_k + 2\frac{\dot{a}}{a}\dot{\delta}_k = 32\pi G\bar{\rho}\delta_k , \tag{10}$$

yielding two solutions $\delta_k \propto t^{\pm 1}$. For large λ the growing radiation- or matter-dominated solutions are in agreement with the heuristic model, but on smaller scales the behavior of each fluid component will depend crucially on whether or not the perturbation exceeds λ_J for that component, and whether the component dominates the expansion.

These considerations factor together into an overall picture as follows. In a hot big-bang, the universe is radiation-dominated until some time $t_{\rm eq}$. During this epoch, perturbations outside the horizon grow as $\delta \propto a^2$. Upon entering the horizon, per-turbations in the radiation and baryons are held up by pressure and fail to grow (in fact they oscillate with near-constant amplitude). The dark matter, being both cold and collisionless, would "like" to grow as $\delta_{\rm dark} \propto a$, and is prevented only by the rapid expansion (dominated by the radiation); the perturbations turn out to grow, but only logarithmically ($\delta_{\rm dark} \propto \ln a$).

At $t_{\rm eq}$, matter begins to dominate the expansion, and $\delta_{\rm dark} \propto a \propto t^{2/3}$. Baryons, however, are still coupled to the radiation; this provides strong pressure support (i.e. a high Jeans mass) so that the perturbations in the baryons, like those in the photons, cannot grow but instead oscillate (as discussed in more detail below). During this epoch the dark matter perturbations can thus grow substantially relative to those in the baryons, a fact which will be of great importance.

Finally, at some time $t_{\rm dec}$, the rate of collisional ionization becomes too low to maintain the ionization of the baryonic fluid, and the nuclei and free electrons combine to form atoms. With few free electrons, the baryons decouple from the photons, the baryonic Jeans mass drops drastically, and baryonic perturbations can subsequently grow as $\delta_{\rm baryon} \propto a \propto t^{2/3}$ on scales $\lambda \gtrsim \lambda_J$ (the latter now being determined by the baryonic pressure rather than that of the photons); on smaller scales they continue to oscillate. In fact, due to their growth during $t_{\rm eq} < t < t_{\rm dec}$, the dark-matter perturbations are now larger than those in the baryons. The baryons can then "fall in" to the existing dark-matter perturbations; thereafter their amplitudes will be equal for $\lambda \gtrsim \lambda_J$, where λ_J is now calculated using the combined baryonic and dark matter density.

The resulting power spectrum of the matter (baryons+dark matter) after decou-pling carries the imprint of this earlier epoch, and this imprint is encapsulated in the "transfer function" T_k:

$$T_k \equiv \frac{\delta_k(z=0)}{\delta_k(z)D(z)}, \tag{11}$$

where $\delta_k(z)$ is the power spectrum at some very early z before which any relevant perturbation had entered the horizon, and $D(z)$ is the "linear growth function" $D(z)$, which is a general expression for the linear growth of perturbations in a homogeneous background, absent effects such as free-streaming or pressure support: $D(z)/D(z_0) = \delta(z)/\delta(z_0)$ for some reference redshift z_0. This is given by[12]

$$D(z) = \frac{5\Omega_m}{2}(1+z_0)g(z)\int^z \frac{1+z'}{g^3(z')}dz',$$

$$g^2(z) = \Omega_m(1+z)^3 + \Omega_\Lambda + (1-\Omega_0-\Omega_\Lambda)(1+z)^2.$$

A full and precise calculation of T_k must be done numerically, and several publicly available codes for doing so exist.[28] However, there are approximations that are sufficiently good for many purposes.[3,16,12] The key features are exhibited in Fig. 1, which shows the transfer function for cold dark matter computed using Eisenstein and Hu.[12] At large scales (small k) it is constant and the power spectrum at late times exactly reflects the "primordial" $n \simeq 1$ power spectrum. At

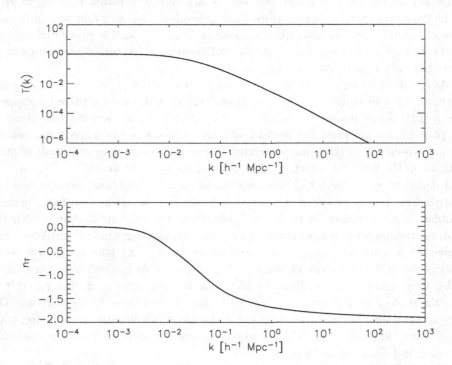

Fig. 1.   Transfer function T for standard cosmological model (top), and power-law index of T (bottom).

large k, corresponding to scales below the horizon size at T_{eq}, T falls as k^{n_T} so that $P(k) \propto k^n T^2(k) \propto k^{n+2n_T}$, with n_T decreasing with k to $n_T \simeq -2$ at the smallest scales (see bottom panel of Fig. 1). If the ratio Ω_B/Ω_{DM} is fairly small, the main effect of baryons on this transfer function is to change the effective density of the dark matter in a scale-dependent way; taking this into account for $\Omega_B/\Omega_{DM} = 0.044/0.226$ and $h = 0.71$ gives T_k as shown in Fig. 1 (top panel).

The power spectrum at decoupling can be directly connected to several important cosmological phenomena. First, the power on large scales continues to grow $\propto a$ and describes the large-scale distribution of matter in the universe, as reflected, for example, in the distribution of galaxies on scales $\gtrsim 10$ Mpc. Second, on smaller scales the perturbations provide the seeds for the nonlinear collapse of the fluid into galaxies and clusters. Third (less important for the universe but more important for our knowledge of it), the perturbations leave a detailed imprint in the CMB. The signature of dark matter in all three of these phenomena is important and testable, and will be discussed in the next several sections.

3. Tests and Constraints from the Microwave Background

Even before the confirmation of the precise thermal spectrum of the CMB by COBE, it was recognized that anisotropies in the observed CMB temperature would provide a snapshot of the density inhomogeneities that existed at the time when the universe first became transparent to photons at t_{dec} (often also called the "recombination time" or "time of last scattering"). A number of excellent reviews of the physics of CMB anisotropies have been written; see, e.g. Hu and Dodelson[15]; here I will qualitatively review the basics, then focus on the role of dark matter.

The observed anisotropies of the CMB may be divided into *primary* anisotropies, which result from density fluctuations at recombination, and *secondary anisotropies* that are imprinted during later cosmological evolution. Primary anisotropies result from three main effects: the gravitational red/blueshift of photons emerging from potential wells, the Doppler shift of photons emitted from a medium with an inhomogeneous velocity field, and the lower-temperature emission of photons from regions that are overdense and hence recombine slightly later. These three effects are often termed, respectively, the "Sachs Wolfe effect", the "Dopper effect", and "intrinsic fluctuations." Since in linear theory the velocity field can be directly related to the density field, all three effects essentially capture the imprint of density inhomogeneities on the last-scattering "surface".

The analysis of the observed CMB generally proceeds by decomposing the temperature $T(\theta, \phi)$ into a sum of spherical harmonics Y_{lm} of amplitude a_{lm}, and computing the angular power spectrum

$$C_l \equiv \frac{1}{2l+1} \sum_{m=-l}^{m=l} |a_{lm}|^2 .$$

Roughly speaking, C_l gives an estimate of the power on angular scales $\approx 180/l$ degrees.

This angular power spectrum contains a multitude of information about the physics and constituents of the universe when the fluctuations were imprinted at $z \approx$ 1100, as well as some information about the subsequent evolution of the universe. The latter come primarily from the angular diameter distance $d_A(z)$ (defined as D/θ, where θ is the angle subtended by an object of physical size D at redshift z), which connects physical scales at the recombination epoch to angular scales in the observed CMB anisotropies. This distance measure contains an integral over redshift that involves the energy densities of all energy components (see Hogg[17] for explicit formulas and other distance measures). Since the physical size of the last scattering surface is known, this gives one constraint on the energy densities that turns out to be quite sensitive to curvature; it is from this measurement[40] that we now know that the universe is geometrically approximately flat ($|\Omega - 1| < 0.05$).

While the angular diameter distance to $z \approx 1100$ sets the overall scaling for C_l, because of the complicated interplay (described in Sec. 2.2) of different components in the evolution of density perturbations between when they enter the horizon and when they are imprinted in the CMB, the measured power spectrum also yields information on, among many other things, the prevalence of dark matter and baryons. Consider first a scenario without dark matter. When fluctuations in the baryon density enter the horizon (or, really, the horizon expands to encompass them), they are pressure supported and oscillate until decoupling, when the CMB is last scattered. At that time the largest scale just able to compress (before rarefying due to pressure) imprints extra power at that scale; this leads to a peak in the power spectrum on a physical scale of the horizon at recombination, or an angular scale of about 0.5 degree. Higher harmonics are represented by a series of peaks at higher-l; the second peak corresponds a scale of maximum rarefaction. (The anisotropies measure the amplitude, rather than the value of the density fluctuations, which is why this shows up as a peak.) Higher-l peaks alternate between compression and rarefaction. With no further effects accounted for, the CMB would resemble a flat line at $l \lesssim 50$ connected to a squared sinusoidal curve at $l \gtrsim 50$. However, damping at high-l due to the finite thickness of the last scattering surface, and the ability of photons to diffuse out of small-scale density wells cause the peaks to fall off steadily in amplitude as l increases.

Collisionless dark matter changes this picture by adding a compression component that has no restoring force, so that the compression modes have greater amplitude than the rarefaction modes. With sufficient dark matter content, there is thus a fall in peak amplitude from the first to second peak, then a *rise* to the third peak. In addition, the additional compressing force makes the peaks somewhat narrower. Such effects provide a signature of dark matter, and their nature can be seen visually by inspecting some of the many reviews of CMB physics in which the cosmological parameters are varied.[15]

A detailed comparison of the WMAP data to a suite of models[44] shows that even with all other parameters left to vary, $\Omega_{DM}h^2 = 0.10 \pm 0.02$ is required, and when only 6 "standard" parameters are free, $\Omega_{DM}h^2 = 0.12^{+0.02}_{-0.02}$.

It is worth commenting here on possible alternatives to the dark matter hypothesis such as that propounded by Milgrom (this volume), in which gravity is modified so as to become stronger at small acceleration scales. Are these now ruled out by the CMB? Perhaps, but it is not as yet entirely clear. It is nearly impossible to see how such models would account for an alternation of peak amplitudes except by extreme luck or contrivance, but the data regarding the third peak is (as of this writing) insufficiently precise to warrant iron-clad conclusions. For the first two peaks, an alternative to dark matter might hope to reproduce the CMB by positing that gravity is unmodified at early times, so the only difference from the standard scenario is the absence of collisionless matter. In this case "no-CDM" models can be generated that provide a quite good qualitative fit to the observed power spectrum.[24] Still, the data on the first two peaks is very high quality, and it does not appear possible to fit it without a substantial contribution by massive neutrinos[44,24] (which would, of course, be non-baryonic dark matter, albeit of a familiar type).

A second test of the presence and importance of dark matter is in the connection between the CMB power spectrum and the power spectrum of galaxies that are the result of structure formation at later times. A given cosmological model provides a precise translation between the power spectrum of initial perturbations, and the power spectrum of matter at late times $z < 4$, to which we now turn.

4. Tests and Constraints from the Ly-α Forest and Distribution of Galaxies

Long before the CMB anisotropies were ever observed, attempts were made to understand the large-scale distribution of matter using the distribution of galaxies, and to use this information to infer the mechanism of structure formation. The primary tool used has, again, been the power spectrum $P(k)$. For galaxies it it useful to relate this to the more easily measured 2-point correlation function

$$\xi(\vec{r}) \equiv \langle \delta(\vec{x})\delta(\vec{x}+\vec{r})\rangle,\qquad(12)$$

which can be written in Fourier-transformed terms in a statistically isotropic universe as:

$$\xi(r) = \frac{V}{(2\pi)^2}\int P(k)\frac{\sin kr}{kr}4\pi k^2 dk,\qquad(13)$$

where V is a volume over which the averaging in Eq. (12) is done. This can be inverted to yield an expression for the power spectrum.

Galaxy surveys can give a measurement of the distribution of the cosmic (mainly dark) matter, but only to a degree that either galaxies (or light from galaxies) trace mass, or that the "bias" in the relation between light and mass is independently known. The study of the power spectrum of galaxies has become quite mature with the completion of substantial parts of the SLOAN and 2dF galaxy surveys.[45,13] These surveys find that bias depends on galaxy type, but is near unity when averaged over all galaxies. Further, both surveys derive a mass power spectrum on comoving scales $\sim$ 20–200 Mpc that is a good match to the prediction obtained by evolving the CMB-measured power spectrum forward in time to the present epoch, using standard cosmological theory. This provides an excellent consistency check on the standard cosmological model. These comparisons are now being used to probe the primordial power spectrum over a wide range of length-scales.[44]

A second way of measuring the mass power spectrum on scales $\lesssim$ 10 Mpc that has recently been developed uses absorption spectra of high-redshift quasars. These spectra are filled with a "forest" of Ly-α absorption features caused by density fluctuations in the highly-ionized IGM at $z \sim$ 1–4. Theoretical arguments[37] and numerical simulations[10] indicate that there is a tight correlation between Ly-α absorption and the density of the absorbing gas, so the correlation function of absorption in a quasar spectrum can be rather directly converted into a 1-d power spectrum of the intergalactic medium.[7] On large enough scales this is expected to closely track the dark matter distribution, and so gives an independent mass power spectrum on rather small scales. Encouragingly, this power spectrum agrees fairly well with the galaxy power spectrum where they overlap, and with the forward-evolved CMB-inferred primordial power spectrum (although there are tantalizing hints that the combination of CMB and Ly-α forest data may call for a non-standard primordial spectrum[40]). Two groups are now undertaking a detailed study of the Ly-α power spectrum using the SLOAN data.

Although the power specra from galaxy surveys and Ly-α absorption are excellent tools for studying dark matter, structure formation, and their interplay, they do not directly address the original purpose for which dark matter was proposed, and in which the behavior of dark matter is least well-understood: its role in galaxy formation and evolution.

For this, we must return to the story of structure formation to the point at which an initially overdense perturbation forms a collapsed dark matter halo that can host a galaxy.

5. Dark Matter and Galaxy Formation

5.1. *Halo Formation*

Once a perturbation grow sufficiently large, it will separate from the background cosmic expansion and collapse to a self-gravitating "halo". Under the assumption of spherical symmetry, this can be seen to happen at a time when the overdensity as calculated by linear theory reaches a critical level Δ that depends (weakly) on the background geometry and the constituents of the fluid; in a universe of only dark matter, $\Delta = 1.69$.

5.2. *The Halo Mass Function*

A truly accurate calculation of the properties of dark matter halos requires direct N-body simulation. Such calculations are well-developed and publicly available codes exist (such as GADGET [42]) that can numerically evolve the cosmological dark matter distribution from a very early time through the age of galaxy formation.

A basic understanding of the distribution function of halo masses can, however, be gained from a simple model pioneered by Press and Schechter[32] that, as it turns out, provides a surprisingly good characterization of the mass function of halos found numerically. In this approach, it is assumed that a structure of mass M collapses when a density perturbation smoothed over that mass scale, as calculated by linear theory, reaches a critical density δ_c. In this picture every overdense region will collapse eventually, and the probability P that a randomly chosen point will be in such a fluctuation of mass M is just

$$P(\delta > \delta_c; M) = \frac{1}{2}\left[1 - \mathrm{erf}\left(\frac{\delta_c}{\sqrt{2}\sigma(M)}\right)\right],$$

because $\sigma(M)$, as described in Sec. 2, is just the width of a Gaussian probability distribution governing the density contrast in regions of mass M. There are then two more steps. First, it is assumed that the *under*dense gas simply accretes onto the collpsed halos, the net effect of which is just to double $P(\delta > \delta_c; M)$ (although just a fudge by Press and Schechter, this is justified in the more rigorous approached mentioned below). Second, it is reasoned that we should "attribute" each mote of dark matter to the most massive collapsed region of which it is part: $2P(\delta > \delta_c; M)$ then describes the fraction of the dark matter that is incorporated into halos of mass $> M$. This is then a cumulative probability function that can be differentiated[30] to yield a mass function $f(M)$, where $f(M)dM$ is the comoving number density of halos of mass between M and $M + dM$:

$$Mf(M) = \frac{\rho_0}{M}\left|\frac{d\ln\sigma}{d\ln M}\right|\sqrt{\frac{2}{\pi}}\exp(-\nu^2/2). \tag{14}$$

Here $\nu \equiv \delta_c/\sigma(M)$, and ρ_0 is the comoving dark matter density.

This function takes the form of a power law with an exponential cutoff at high-mass, and as such provides the potential for accurately fitting luminosity functions of galaxies, which tend to have this "Schechter function" form. However, the observed luminosity function exhibits an approximate behvior of $f(M) \propto M^{-1}$, whereas Eq. (14) gives $f(M) \propto M^{(n-9)/6}$, using Eq. (6). Since on the scale of small galaxies, $n < -2$, the predicted number of very small halos greatly exceeds the number of observed faint galaxies. This points to the "satellite problem" that has been greatly discussed as one of the challenges for the CDM paradigm in galaxy formation; but of course there is no assurance that small halos should necessarily all form galaxies around them.[c]

Although the Press-Schechter approach often suffices for everyday use, there are more advanced treatments[5,38] (still short of direct simulation) that provide greater accuracy, and can yield additional information such as (a statistical description of) the merger history of halos.

5.3. *Halo Profiles*

Numerical studies[27] have shown that within the standard CDM model, halos collapse to a nearly universal form with a spherically-averaged density profile of

$$\rho(r) = \frac{\rho_c}{r[1 + (r/r_c)^2]} , \tag{15}$$

where ρ_c is the central density and r_c is a "core radius". These parameters can be expressed in various combinations of the virial radius r_v (the radius within which the mean density is, say, $200\bar{\rho}$), the concentration parameter $c \equiv r_v/r_c$, and the halo mass M_v within r_v. The physical reason that the halo density profile takes this universal form is unclear, though some attempts to derive it analytically have been made.

The profile of Eq. (15) has some features of interest for the theory of galaxy formation. First, the corresponding circular velocity profile is approximately flat for $r \sim r_c$ (as one would hope, in order to explain the flat rotation profile of observed galaxies), but slowly falls off at large-r. Second, it contains a rather steep central cusp of $\rho \propto r^{-1}$. The exact slope of this central cusp has been a matter of some debate and consternation,[27,23,35] because a number of (particularly small, low surface-brightness) galaxies appear to have rotation curves inconsistent with such a central cusp.[11,43] Although a very active area, the issue of whether the fault for this discrepancy lies in the CDM predictions, the CDM model itself, or in the accuracy of the observations is at present still rather unresolved.

5.4. *Angular Momentum*

Also the key for galaxy formation is the angular momentum $\vec{J}$ of the collapsing halo, the magnitude of which can be expressed in dimensionless terms as

$$\lambda \equiv \frac{|\vec{J}|E^{1/2}}{GM_v^{5/2}} , \tag{16}$$

where E is the binding energy of the object[30]; $\lambda = 1$ would correspond to angular momentum fully supporting the halo against collapse. In numerical simulations,[46]

---

[c]The predicted dark matter halo distribution and the galaxy luminosity functions are also discrepant at the *high*-mass end; this can be seen very clearly in clusters which do *not* resemble gigantic galaxies.

$\ln \lambda$ is found to be distributed normally among halos, with $\langle \ln \lambda \rangle \approx -3.2$ and $\sigma \approx 0.5$–0.6. The angular momentum is thought to result from tidal torquing by nearby perturbations during collapse, but may also have contributions from the accretion of smaller halos during subsequent evolution.[48]

5.5. *From a Dark Halo to a Galaxy*

The canonical description of galaxy formation after the collapse of a halo goes as follows. The gas in the halo, unlike the dark matter, can dissipate energy by cooling. This leads to a contraction of the gaseous halo to a radius $\sim \lambda r_v$. At this point, the halo is supported by angular momentum and cannot contract further in the plane perpendicular to the total angular momentum vector. This leads to the formation of a thin, axisymmetric disk with a density profile that is determined in part by the initial angular momentum distribution of the gaseous halo. During the collapse process, the increased baryonic density at small radii tends to contract the dark halo, potentially enhancing even further the steep density profile predicted by simulations.

What happens next to the disk depends upon what instabilities exist in the disk density structure. A disk without a dark halo is highly unstable to the formation of a bar — indeed this was another early argument for dark halos — and may form one even in a halo's presence. On a local level, density perturbations in the disk are stabilized by shearing in the disk (which manifests in an r−dependence of the angular rotation speed Ω) and thermal pressure. They can grow only when the dynamical time of some region characterized by a surface density Σ is shorter than both the sound-crossing time and the shearing time. This gives rise to the Toomre Q−parameter[4]

$$Q \equiv \frac{v_s \kappa}{\pi G \Sigma}$$

where $\kappa^2 = \frac{d}{dr}\Omega^2 + 4\Omega^2$, and v_s is the sound speed. For $Q > 1$, local perturbations are unstable against growth, and star formation can presumably proceed.

5.6. *Current Status of Galaxy Formation Theory*

The simple picture just outlined neglects an enormous set of complicated physical processes that play a part in galaxy formation. Two particularly important ones are, first, that halos accrete and collide with other halos and second, that energy released from star formation affects the physics of the gas.

A great amount of work has been performed to attempt to treat these and other complicated processes to assemble a reasonably comprehensive picture of galaxy formation that can be compared to galaxy observations. There are two basic approaches in this project. In the numerical approach, numerical simulations including gas dynamics are evolved from an early time to produce an *ab-initio* calculation of galaxy properties today. In the "semi-analytic" approach, in simplified prescriptions for physics such as gas cooling, star formation, and feedback from stellar energy release are added to already completed dark-matter-only simulations (or, in simpler models, extensions of the Press-Schechter approach) to produce a set of statistical predictions for galaxy properties.[d] In both approaches predictions can be made for the luminosity and mass function of galaxies, the global star formation history, and other observables.

Overall, both programs have met with a great deal of success. Many of the observed bulk properties of galaxies, as well as trends in those properties with time or galaxy mass, are reasonably reproduced. This would by no means be assured in any alternative to the standard CDM model. Nonetheless, there are several outstanding difficulties in the details comparison of CDM galaxy formation theory to observations that are sufficiently severe that they have led some theorists to contemplate modifying the dark matter properties or abandoning the notion of dark matter altogether.

5.7. *Outstanding Problems, and Alternatives to (Cold) Dark Matter*

The first possible problem was mentioned in Sec. 5.2: CDM theory predicts a number of low-mass halos that is much larger than the number of low-mass galaxies we observe. This had long been noticed in semi-analytic galaxy formation model,[19,39] in which feedback was invoked to reduce the small-halo abundance. It was made more acute when simulation groups produced dark-matter simulations meant to resemble the Milky Way halo and discovered $\sim$ 100–1000 simulated dark-matter satellites, compared to only $\sim$ 10 detected satellite galaxies.[20,22] This problem, however, has a number of quite plausible solutions — there is no particular reason to believe that very small halos should have stars, and good reasons to believe they should not: cosmic radiation after reionization could evaporate them, and feedback could blow away their star-forming gas. In addition, the "missing" dark matter subhalos may now be showing up observationally in the form of flux-ratio anomalies in multiply-images lensing systems[8] which indicate dark substructure in galaxies, with approximately the density predicted in CDM theory.

The second potential problem was mentioned in Sec. 5.3: dark matter halos are expected to have a steep $\rho \propto r^{-1}$ density cusp in their centers, yet observed dark matter density profiles, as inferred from the dynamics of the central gas and stars, tend to be better fit by a model with a constant-density core, and are often outright incompatible with an r^{-1} cusp.[11] The status of this problem is still not entirely clear, though the observers and simulators are rapidly improving their results and understanding to what degree there is conflict.

The third potential problem concerns the halo angular momentum discussed in Sec. 5.4. Early calculations showed that $\lambda \sim 0.05$, typical for galaxy halos, would

---

[d]Due to limited resolution, numerical simulations also must add parametrized prescriptions for processes such as star formation.

lead to a disk size comparable to observed spiral galaxy disks *if* the angular momentum of the gas was strictly conserved.[50] This nice general idea, however, breaks down when implemented in more detail. First, numerical simulations of galaxy formation find disks that tend to be far too small. This is believed to occur due to tansfer of angular momentum from the gas to the dark matter, but it is as yet unclear whether this is a correct physical effect or an artifact of limited numerical resolution.[26] Semi-analytic models also have difficulties, in that the net amount of angular momentum in halos is approximately right (*if* conserved) but the distribution of angular momentum in simulated dark halos, if applied to the gas and conserved parcel-by-parcel, does not lead to an exponential disk.[6] Like the density profile, this problem is a subject of significant current attention.[34]

A final potential problem, slightly harder to crisply define, concerns the systematic properties of galaxies. Despite the array of complicated and stochastic effects expected to be integral to galaxy formation (e.g., merger, starbursts, galactic feedback and winds, environmental effects, random formation times, etc.), the properties of spiral galaxies appear to be remarkably regular. For example, the Tully-Fisher relation between luminosity and asymptotic rotation speed is compatible with being nearly exact — i.e. the scatter in the relation could plausibly be entirely observational error.[47] In more detail, the simple relation proposed by Milgrom (this volume) as a formulation of modified gravity fits the systematics of galaxies extraordinarily well — given the observed gas/star density profile, the observed rotation curve can be accurately predicted using only at most one free parameter.[36] If CDM theory is correct this requires a very tight (and probably not-quite-understood) coupling between the visible and dark matter.

This rash of problems initially provoked a number of proposed modifications of dark matter, e.g. to make it slightly warm,[14,49] or self-interacting.[41] The idea of all of these was to reduce the small-scale structure — whether in halo cores or in tiny subhalos. These models appear to have fallen largely out of favor, partially due to the ameliorization of the "subhalo problem", and partly because modifying the cores of dwarf galaxies (where rotation curves are well-measured) without significantly altering the density profile in the cores of clusters is difficult.[9,21] Also, because it probes small scales, the Ly-α can place direct constraints on dark matter itself; if dark matter particles were light enough that they were not completely cold, the free streming of the particles would erase small-scale power in the density field. Current measurements from the Ly-α forest put a lower bound of ~ 750 eV on the dark matter particle mass.[25]

The final, rather radical alternative to CDM theory that bears mentioning is MOND (see this volume), conceived as a modification of gravity that would obviate the need for dark matter. As mentioned above, MOND has great success in accounting for the observed dynamics of galaxies and aacounting for the systematics of galaxy properties. However, because it is not a full theory, it is far less predictive than CDM: the CMB anisotropies, large-scale galaxy distribution, galaxy mass function, etc., cannot be reliably calculated and tested, and in some cases where MOND does make some firm predictions (cluster dynamics, absence of substructure in galaxies, etc.) it runs somewhat afoul of observations.[1] The success of MOND seems to be pointing to something, but whether or not it points to the need to modify gravity and banish dark matter, I leave it to the reader to decide.

6. Conclusions

Although much has happened in astronomy and astrophysics since the 1986 Jerusalem Winter school, several things remain the same. First, the fundamental nature of the dark matter is, as of this writing, still completely unknown. Its elucidation ranks as one of the foremost tasks in astrophysics, and given the enormous effort currently being put forth by many observational and experimental groups, we have a reasonable hope that it may be forthcoming in the relatively near future. Second, although many techniques of studying the dark matter have become somewhat more sophisticated in detail, they are the same in their basic structure, and rely on the same basic physics, as when the following lectures were compiled nearly two decades ago. The student will, therefore, find a great deal in the preceding chapters that will help build a foundation for understanding a range of topics in modern astrophysics and cosmology.

References

1. Aguirre, A., astro-ph/0310572
2. Bardeen, J. M., *Phys. Rev.* **D22**, 1882 (1980).
3. Bardeen, J. M., J. R. Bond, N. Kaiser, and A. S. Szalay *Astrophys. J.*, **304**, 15 (1986).
4. Binney, J. and S. Tremaine, *Galactic Dynamics*, Princeton, NJ: Princeton University Press (1987).
5. Bond, J. R., S. Cole, G. Efstathiou, and N. Kaiser, *Astrophys. J.*, **379**, 440 (1991).
6. Bullock, J. S., A. Dekel, T. S. Kolatt, A. V. Kravtsov, A. A. Klypin, C. Porciani, and J. R. Primack, *Astrophys. J.*, **555**, 240 (2001).
7. Croft, R. A. C., D. H. Weinberg, M. Pettini, L. Hernquist, and N. Katz, *Astrophys. J.*, **520**, 1 (1999).
8. Dalal, N. and C. S. Kochanek, *Astrophys. J.*, **572**, 25 (2002).
9. Dalcanton, J. J. and C. J. Hogan, *Astrophys. J.*, **561**, 35 (2001).
10. Davé, R., L. Hernquist, N. Katz, and D. H. Weinberg, *Astrophys. J.*, **511**, 521 (1999).
11. de Blok, W. J. G., in *Galaxy Evolution: Theory and Observations* (eds. V. Avila-Reese, C. Firmani, C. S. Frenk and C. Allen), Revisita Mexicana de Astronomy Astrofisica (Serie de Conferencias) Vol. 17, pp. 17–18 (2003).
12. Eisenstein, D. J. and W. Hu, *Astrophys. J.*, **511**, 5 (1999).
13. Hawkins, E. *et al.*, *Mon. Not. R. Astron. Soc.*, **346**, 78 (2003).
14. Hogan, C. J. and J. J. Dalcanton, *Phys. Rev.*, **D62**, 063511 (2000).
15. Hu, W. and S. Dodelson, *Ann. Rev. Astron. Astrophys.*, **40**, 171 (2002).
16. Hu, W. and N. Sugiyama, *Astrophys. J.*, **444**, 489 (1995).
17. Hogg, D., astro-ph/9905116
18. Hu, W., U. Seljak, M. White, and M. Zaldarriaga, *Phys. Rev.*, **D57**, 3290 (1998).
19. Kauffmann, G., S. D. M. White, and B. Guiderdoni, *Mon. Not. R. Astron. Soc.*, **264**, 201 (1993).

20. Klypin, A., A. V. Kravtsov, O. Valenzuela, and F. Prada, *Astrophys. J.*, **522**, 82 (1999).
21. Miralda-Escudé, J., *Astrophys. J.*, **564**, 60 (2002).
22. Moore, B., S. Ghigna, F. Governato, G. Lake, T. Quinn, J. Stadel, and P. Tozzi, *Astrophys. J.*, **524**, L19 (1999).
23. Moore, B., T. Quinn, F. Governato, J. Stadel, and G. Lake, *Mon. Not. R. Astron. Soc.*, **310**, 1147 (1999).
24. McGaugh, S., astro-ph/0312570 (2003).
25. Narayanan, V. K., D. N. Spergel, R. Davé, and C. Ma, *Astrophys. J.*, **543**, L103 (2000).
26. Navarro, J. F. and M. Steinmetz, *Astrophys. J.*, **478**, 13 (1997).
27. Navarro, J. F., C. S. Frenk, and S. D. M. White, *Astrophys. J.*, **490**, 493 (1997).
28. Zaldarriaga, M. and U. Seljak, *Astrophys. J. Suppl. Ser.*, **129**, 431 (2000).
29. Padmanabhan, T., *Structure Formation in the Universe*, Cambridge, UK: Cambridge University Press (1993).
30. Peacock, J. A., *Cosmological Physics*, Cambridge, UK: Cambridge University Press (1999).
31. Peebles, P. J. E., *Principles of Physical Cosmology*, Princeton, NJ: Princeton University Press (1993).
32. Press, W. H. and P. Schechter, *Astrophys. J.*, **187**, 425 (1974).
33. Press, W. H. and E. T. Vishniac, *Astrophys. J.*, **239**, 1 (1980).
34. Primack, J. R., astro-ph/0312547.
35. Primack, J. R., astro-ph/0312549.
36. Sanders, R. H. and S. S. McGaugh, *Ann. Rev. Astron. Astrophys.*, **40**, 263 (2002).
37. Schaye, J., *Astrophys. J.*, **559**, 507 (2001).
38. Sheth, R. K., H. J. Mo, and G. Tormen, *Mon. Not. R. Astron. Soc.*, **323**, 1 (2001).
39. Somerville, R. S. and J. R. Primack, *Mon. Not. R. Astron. Soc.*, **310**, 1087 (1999).
40. Spergel, D. N., *et al.*, *Astrophys. J. Suppl. Ser.*, **148**, 175 (2003).
41. Spergel, D. N. and P. J. Steinhardt, *Phys. Rev. Lett.*, **84**, 3760 (2000).
42. Springel, V., N. Yoshida, and S. D. M. White, *New Astronomy.*, **6**, 79 (2001).
43. Swaters, R. A., B. F. Madore, F. C. van den Bosch, and M. Balcells, *Astrophys. J.*, **583**, 732 (2003).
44. Tegmark, M., *et al.*, astro-ph/0310723 (2003).
45. Tegmark, M., *et al.*, astro-ph/0310725 (2003).
46. van den Bosch, F. C., T. Abel, R. A. C. Croft, L. Hernquist, and S. D. M. White, *Astrophys. J.*, **576**, 21 (2002).
47. Verheijen, M., *Astrophys. J. Suppl. Ser.*, **269**, 671 (1999).
48. Vitvitska, M., A. A. Klypin, A. V. Kravtsov, R. H. Wechsler, J. R. Primack, and J. S. Bullock, *Astrophys. J.*, **581**, 799 (2002).
49. White, M. and R. A. C. Croft, *Astrophys. J.*, **539**, 497 (2000).
50. White, S. D. M. and M. J. Rees, *Mon. Not. R. Astron. Soc.*, **183**, 341 (1978).

Printed in the United States
By Bookmasters